# Environmentally Friendly Manufacturing of Food Additives and Functional Ingredients: Volume II

# Environmentally Friendly Manufacturing of Food Additives and Functional Ingredients: Volume II

Guest Editors

**Wenjie Sui**
**Chanchan Sun**

MDPI

Basel • Beijing • Wuhan • Barcelona • Belgrade • Novi Sad • Cluj • Manchester

*Guest Editors*
Wenjie Sui
College of Food Science and Engineering
Tianjin University of Science and Technology
Tianjin
China

Chanchan Sun
School of Life Sciences
Yantai University
Yantai
China

*Editorial Office*
MDPI AG
Grosspeteranlage 5
4052 Basel, Switzerland

This is a reprint of the Special Issue, published open access by the journal *Foods* (ISSN 2304-8158), freely accessible at: https://www.mdpi.com/journal/foods/special_issues/B17J987832.

For citation purposes, cite each article independently as indicated on the article page online and as indicated below:

| Lastname, A.A.; Lastname, B.B. Article Title. *Journal Name* **Year**, *Volume Number*, Page Range. |
|---|

**ISBN 978-3-7258-7645-7 (Hbk)**
**ISBN 978-3-7258-7646-4 (PDF)**
**https://doi.org/10.3390/books978-3-7258-7646-4**

# Contents

# Preface

In an era marked by increasing global concerns regarding food safety, environmental sustainability, and public health, the development of food additives and functional ingredients that balance performance, environmental compatibility, and nutritional value has become an urgent priority for the food science community. Against this backdrop, we are pleased to present the Reprint of the Special Issue "Environmentally Friendly Manufacturing of Food Additives and Functional Ingredients (Volume II)" in *Foods*, a follow-up to our highly regarded first volume that continues to highlight cutting-edge research at the forefront of sustainable food innovation.

The central focus of this Special Issue is the re-examination of the production and application of food additives and functional ingredients through an environmental perspective—emphasizing renewable resources, low-carbon processes, and biodegradable outcomes, while maintaining or enhancing the sensory, stability, and health-promoting properties of food products. Its scope is intentionally broad, encompassing topics such as the extraction and modification of natural functional components (e.g., polysaccharides, proteins, and plant-derived bioactives), the development of sustainable alternatives to synthetic additives (including colorants, fat replacers, and preservatives), the optimization of environmentally efficient manufacturing technologies (such as extrusion, microwave processing, and pH-controlled extraction), and the evaluation of the safety, functionality, and environmental impact of novel ingredients.

The motivation for this Special Issue arises from two critical imperatives: first, the necessity to address the environmental limitations of conventional food additive production, which frequently depends on non-renewable resources, generates substantial waste, or employs hazardous chemical processes; second, the increasing consumer demand for "clean label" foods—products devoid of artificial additives, abundant in natural nutrients, and consistent with sustainable consumption principles. By compiling this collection, we seek to bridge the gap between academic research and industrial application, offering a platform for researchers, engineers, and practitioners to disseminate innovative methodologies, evaluate practical solutions, and foster collaborative efforts to advance the food industry.

This Special Issue targets a diverse audience committed to advancing sustainable food science. It addresses academic researchers specializing in food chemistry, food engineering, and functional nutrition, providing insights into emerging materials and technologies; food manufacturers and product developers pursuing scalable, environmentally sustainable alternatives to conventional additives; policymakers and regulatory experts focused on establishing standards for sustainable food ingredients; and students and early-career professionals seeking to engage with contemporary developments in sustainable food innovation. We anticipate that the research presented in this Special Issue will not only contribute to scientific discourse, but also facilitate measurable progress toward a more sustainable, health-conscious, and resilient food system—one in which environmental stewardship and food quality are mutually reinforcing.

We express our sincere appreciation to all authors for their valuable contributions, to the reviewers for their rigorous peer review, and to the editorial team of *Foods* for their steadfast support. We hope this Special Issue will serve as a catalyst for ongoing innovation, collaboration, and impact the field of environmentally sustainable food additive and functional ingredient production.

**Wenjie Sui and Chanchan Sun**
*Guest Editors*

foods

*Article*

# Fabricating Pea Protein Micro-Gel-Stabilized Pickering Emulsion as Saturated Fat Replacement in Ice Cream

**Xv Qin [1,†], Yaxian Guo [1,†], Xiaoqing Zhao [1], Bin Liang [2,*], Chanchan Sun [1,*], Xiulian Li [3] and Changjian Ji [4]**

1 College of Life Sciences, Yantai University, Yantai 264005, China; 15063757029@163.com (X.Q.); gyx19991009@163.com (Y.G.); zxqytu1011@163.com (X.Z.)
2 College of Food Engineering, Ludong University, Yantai 264025, China
3 School of Pharmacy, Binzhou Medical University, Yantai 264003, China; ouclixiulian@163.com
4 Department of Physics and Electronic Engineering, Qilu Normal University, Jinan 250200, China; jichangjiansdejcj@qlnu.edu.cn
* Correspondence: liangbin1989311@163.com (B.L.); sunchan88@126.com (C.S.)
† These authors contributed to the work equally and should be regarded as co-first authors.

**Abstract:** Unsaturated fat replacement should be used to reduce the use of saturated fat and *trans* fatty acids in the diet. In this study, pea protein micro-gels (PPMs) with different structures were prepared by microparticulation at pH 4.0–7.0 and named as PPM (pH 4.0), PPM (pH 4.5), PPM (pH 5.0), PPM (pH 5.5), PPM (pH 6.0), PPM (pH 6.5), and PPM (pH 7.0). Pea protein was used as a control to evaluate the structure and interfacial properties of PPMs by particle size distribution, Fourier transform infrared spectroscopy (FTIR), free sulfhydryl group content, and emulsifying property. PPM (pH 7.0) was suitable for application in O/W emulsion stabilization because of its proper particle size, more flexible structure, high emulsifying activity index (EAI) and emulsifying stability index (ESI). The Pickering emulsion stabilized by PPM (pH 7.0) had a uniform oil droplet distribution and similar rheological properties to cream, so it can be used as a saturated fat replacement in the manufacture of ice cream. Saturated fat was partially replaced at different levels of 0%, 20%, 40%, 60%, 80%, and 100%, which were respectively named as PR0, PR20, PR40, PR60, PR80, and PR100. The rheological properties, physicochemical indexes, and sensory properties of low-saturated fat ice cream show that PPM (pH 7.0)-stabilized emulsion can be used to substitute 60% cream to manufacture low-saturated fat ice cream that has high structural stability and similar melting properties, overrun, and sensory properties to PR0. The article shows that it is feasible to prepare low-saturated fat ice cream with PPM (pH 7.0)-stabilized Pickering emulsion, which can not only maintain the fatty acid profile of the corn oil used, but also possess a solid-like structure. Its application is of positive significance for the development of nutritious and healthy foods and the reduction of chronic disease incidence.

**Keywords:** pea protein; micro-gel; O/W emulsion; saturated fat replacement; low-saturated fat ice cream

## 1. Introduction

Chronic diseases, such as cardiovascular disease, have become the most serious diseases threatening humans in the world today, seriously endangering people's physical health. [1]. Many studies have shown that the intake of saturated fats and *trans* fatty acids increases the risk of chronic disease, whereas the intake of unsaturated fats decreases it [2,3]. Therefore, in recent years, the nutritional guidelines by various government agencies have recommended that unsaturated fat replacements could be used to reduce the use of saturated fat and *trans* fatty acids in the diet, so as to achieve a proportion of solid fat in total daily calorie intake of less than 10% [4].

Ice cream is a common oil-in-water system food. It is very popular among consumers because of its delicate, smooth taste and rich, mellow flavor. At present, ice cream processing generally continues the previous traditional ice cream processing methods, with water, milk, cream, sucrose, etc. as the main raw materials. It is characterized by high sugar, calories, and saturated fatty acid. Although consumers are already aware that excessive intake of saturated fat is not conducive to human health, ice cream requires saturated fat to provide structure and a creamy mouthfeel [5]. Therefore, researchers have begun to investigate different techniques for reducing saturated fat to ensure that healthier, nutritious foods are obtained without degrading or destroying the structural and organoleptic properties of the food [6].

Ice cream is a favorite casual dessert, and the market for low-saturated fat ice cream is growing rapidly as consumers focus on healthy eating. However, simply reducing the saturated fat content of ice cream can affect many properties such as viscosity, hardness, melting, and flavor. The difficulty in low-saturated fat ice cream is how to reduce the saturated fat content of ice cream without affecting its sensory properties [7]. Currently, the main approach to reduce saturated fat in food is to use saturated fat replacements (saturated fat mimics and saturated fat replacers like protein-based fat replacers, carbohydrate-based fat replacers, lipid-based fat replacers, and complex fat replacers) to substitute some or all of the saturated fat to ensure sensory properties while reducing the calories [8]. However, these saturated fat replacements have not been commercialized due to a number of problems. The ester bonds of saturated fat replacements cannot be hydrolyzed by lipases and therefore cannot be digested and absorbed by the body [9]. Thus, high intake of saturated fat replacements can lead to many intestinal problems. The application of protein matrix and carbohydrate matrix fat mimics would lead to loss of fat-soluble flavor substances, shortening the shelf life [10,11]. Therefore, the development of a novel saturated fat replacement is of positive significance for the development of nutritious and healthy foods.

In recent years, consumers have gradually realized that saturated fatty acid intake increases the incidence of chronic diseases, so the demand for low-saturated fat food increases year-by-year. Protein-stabilized O/W emulsions, including pea protein-, soybean protein-, wheat gliadin- and corn gliadin-stabilized emulsions, have attracted more and more attention as new saturated fat replacements [12]. On the one hand, protein-stabilized emulsions enable the solidification of vegetable oils. On the other hand, vegetable oil contained in plant protein-stabilized emulsions contains more unsaturated fat esters than fat, which is more nutritious [13]. However, protein-stabilized conventional emulsions are sensitive to environmental factors, such as acid and heat, and tend to destabilize during processing and storage [14]. Therefore, solving this destabilization problem by modifying the molecular structure of proteins is of increasing interest to the food industry.

The protein micro-gel particles produced by microparticulation require heating of the protein to form a gel structure while/later applying high shear to disrupt the gel network into particles [15]. Protein micro-gel particles tend to have great interfacial properties because of the increasing hydrophobicity of the protein surface during microparticulation [16]. Pickering emulsion stabilized by protein micro-gel particles has advantages such as good stability and biocompatibility [17]. Thus, we hypothesized that a protein micro-gel particle-stabilized Pickering emulsion could be used to replace saturated fat in ice cream.

Pea protein has antioxidant, anti-hypertension, anti-inflammatory, cholesterol reduction, and regulation of intestinal bacterial activity, and it has hypo-sensitization [18]. However, the poor solubility, emulsification, and emulsification stability of pea protein limit its application in food processing [19]. Thus, in this study, pea protein was microparticulated to improve its processing performance, such as emulsification. Different structural and physicochemical properties of pea protein micro-gel particles were obtained by adjusting the preparation pH during microparticulation. Then, an O/W emulsion stabilized by pea protein micro-gel particles was designed to substitute saturated fat and was used as saturated fat replacement in low-saturated fat ice cream. And the saturated fat in ice cream was partially replaced at different rates of 20%, 40%, 60%, 80%, and 100%. The physicochemical properties and rheological properties of full-saturated fat and low-saturated fat ice cream were investigated to find the most suitable replacement ratio.

## 2. Materials and Methods

### 2.1. Materials

Pea protein isolate (PPI, containing 85% protein) was obtained from Youcheng Biotechnology Co., Ltd. (Yantai, China). Corn oil was obtained from a local supermarket (Shandong, China). Sodium dodecyl sulfate was obtained from Shanghai MacLean Biochemical Technology Co., Ltd. (Shanghai, China). Skim milk powder was purchased from Dulbert Erie Dairy Co., LTD. Dilute cream was obtained from Qingdao Nestle Co., Ltd. (Qingdao, China) Anhydrous citric acid was obtained from Weifang Yingxuan Industrial Co., Ltd. (Weifang, China) Sodium alginate was obtained from Lianyungang Tiantian Seaweed Industry Co., Ltd. (Lianyungang, China) Sucralose was obtained from Shandong Sanhe Vicin Biotechnology Co., Ltd. (Tai'an, China)

### 2.2. Preparation of Pea Protein Micro-Gel Particles

PPI (18 g/100 mL) dispersion was magnetically stirred for 2 h, and the initial pH of PPI dispersion was 7.22. Then, the pH of PPI dispersion was adjusted from 4.0 to 7.0 using 1 M HCl solution. Then, the PPI dispersions at pH 4.0 to 7.0 were respectively heated at 90 °C for 30 min. The PPI gels were prepared by cooling the heated PPI dispersions to room temperature and then refrigerating at 4 °C for 12 h. Then, the PPI gels were homogenized in a T18 Ultra-Turrax (IKA, Staufen, Germany) at 8000 rpm for 3 min to obtain PPI micro-gel particles. They were named pea protein micro-gel (PPM) (pH 4.0), PPM (pH 5.5), PPM (pH 5.0), PPM (pH 6.0), PPM (pH 6.5), and PPM (pH 7.0), respectively.

### 2.3. Protein Structure Characteristics

#### 2.3.1. Particle Size Distribution of PPMs

Particle size distribution of the protein micro-gel was measured by a Battersize2000 particle size analyzer (Better, Dandong, China). The samples were added to a stirred measuring cell containing 800 mL of deionized water at room temperature, with deionized water as the dispersing medium. And the light shading percentage was between 10% and 15% in order to avoid the multiple scattering effects of proteins and to ensure the accuracy of the measurement. The relative refractive indices of the micro-gel particles and the dispersing medium (deionized water) were 1.52 and 1.33, respectively. Each sample with different treatments was measured three times and the average value was taken.

#### 2.3.2. Free Sulfhydryl Determination of PPI and PPMs

Free sulfhydryl ($SH_F$) was measured according to Yang et al. [20], with minor modifications. Each sample was diluted to 10 mg/mL and then mixed with 3 mL of Tris-glycine buffer (0.086 M Tris-glycine, 0.004 M EDTA, pH 8.0) and 0.03 mL of Ellman's reagent (4.0 mg/mL DTNB) in the same buffer. After fully dissolving, the mixtures were centrifuged at 8000 rpm for 10 min. A Tris-glycine buffer and Ellman reagent were added into the supernatant and mixed evenly. A412 was determined by UV spectrophotometer (UV-2500, Shimadzu, Shanghai, China) after 25 min. In the experiment, the solution with the Ellman reagent without protein was used as the control. $SH_F$ content was calculated according to Equation (1), written as

$$SH_F(\mu mol/g) = \frac{73.53A_{412} \times D}{C} \quad (1)$$

where $D$ is dilution ratio and $C$ is the protein concentration (mg/mL).

#### 2.3.3. Fourier Transform Infrared Spectroscopy of PPI and PPMs

The wavenumber range of the iS10 FTIR spectrometer (Thermo Nicolet, Madison, WI, USA) of Nicole Power was 4000–400 $cm^{-1}$. The resolution of the spectrometer was 4 $cm^{-1}$. The signal to noise ratio was 50,000:1, and the scanning count was 32 times [21].

### 2.4. Emulsifying Properties of PPI and PPMs

Emulsifying activity index (EAI) and emulsion stability index (ESI) were measured according to Figueroa-González et al. [22], with minor modifications. A total of 2 mL corn oil and 8 mL protein solution were mixed. Then, the mixture was homogenized by using a T18 Ultra-Turrax (IKA, Staufen, Germany) at 10,000 rpm for 3 min. Immediately, 50 μL of the emulsion was diluted with 2.45 mL 0.1% SDS solution. The A0 and A10 of the diluted emulsion after 0 min and 10 min of resting at 500 nm were recorded. Then, the EAI and ESI values were calculated using Equations (2) and (3), written as

$$\text{EAI } (\text{m}^2/\text{g}) = \frac{2 \times 2.303 \times A_0 \times \text{DF}}{\text{C} \times \varphi \times 10000} \tag{2}$$

$$\text{ESI } (\%) = \frac{A_{10}}{A_0} \times 100\% \tag{3}$$

where φ is the volume fraction of corn oil ($v/v$) (φ = 0.20), DF denotes dilution ratio (200), and C is the concentration of emulsifier (g/mL).

### 2.5. Emulsion Preparation and Characterization

#### 2.5.1. Emulsion Preparation

A 60 wt% oil phase and 40 wt% PPI or PPM (pH 7.0) were mixed to prepare the Pickering emulsion. The mixture was sheared at 10,000 rpm for 5 min by using the T18 Ultra-Turrax (IKA, Staufen, Germany).

#### 2.5.2. Microstructure Observation

The microstructure of the emulsion droplets was observed using LSM800 CLSM (Zeiss, Bartenwueburg, Germany). A total of 20 μL of 0.1% ($w/v$) Nile red solution was used as a fluorescent dye of oil in the 1 mL emulsion, with excitation and emission wavelengths at 633 nm and 488 nm, respectively [23].

#### 2.5.3. Rheological Properties

Dynamic Oscillatory Measurements

The emulsion was characterized by a MARS 60 rheometer (Haake, Karlsruhe, Germany) at 25 °C [24]. The storage modulus $G'$ (Pa), loss modulus $G''$ (Pa), and tan δ of the emulsions, with frequency ranging from 0.01 to 10 Hz, were measured using a clamp with a conical (25 mm) diameter and a distance of 1 mm between parallel plates. $G'$, $G''$, and the angular frequency $\omega$ were fitted using Equations (4) and (5), written as

$$G' = K'(\omega)^{n'} \tag{4}$$

$$G'' = K''(\omega)^{n''} \tag{5}$$

where $K'$ (Pa·s$^{n'}$) and $K''$ (Pa·s$^{n''}$) are constants and $n'$ and $n''$ may be referred to as the frequency exponents.

Steady-State Shear Properties

The steady-state shear properties of the PPM (pH 7.0)-stabilized emulsion were measured by the MARS 60 rheometer (Haake, Karlsruhe, Germany) at 25 °C. Shear rate was set to increase from 1 s$^{-1}$ to 300 s$^{-1}$ in 400 s [25]. Shear rate ($\gamma$, s$^{-1}$) and apparent viscosity (η, Pa·s) were fitted by a power-law model (Equation (6)). Then, the consistency coefficient $K$ (Pa·s$^n$) and flow behavior index $n$ were obtained [10].

$$\eta = K\gamma^{n-1} \tag{6}$$

2.5.4. In Vitro Digestion

In Vitro Digestion Model

The in vitro digestion model, including three digestion stages of saliva, stomach, and intestine, was created [26,27]. The formulations of the stimulated saliva fluid (SSF), stimulated gastric fluid (SGF), and stimulated intestinal fluid (SIF) are shown in Table S1.

In general, for the oral phase, 5 g of the emulsions were mixed with 4 mL SSF electrolyte stock solution (contained 3.75 mg/mL mucin). Then, 0.975 mL of deionized water and 25 μL $CaCl_2$ solution (0.3 mol/L) were added to the mixture. After the pH was adjusted to 7.0, the mixtures were then stirred at 200 rpm for 2 min (37 °C).

For the gastric phase, 7.5 mL SGF, 1.6 mL porcine pepsin solution (25,000 U/mL), 5 μL of 0.3 mol/L $CaCl_2$ solution, and 148 μL of 6 mol/L HCl solution were added into 10 mL of the above mixture while stirring at 300 rpm for 2 h at 37 °C.

For the intestinal phase, after the pH was adjusted to 7.0 using a 1 mol/L NaOH solution, the gastric chyme for each sample was mixed with a 18.5 mL SIF and 40 μL of 0.3 mol/L $CaCl_2$ solution. The mixtures were stirred at 200 rpm for 2 h at 37 °C.

Particle Size Distribution and Microstructure of Emulsions during In Vitro Digestion

Particle size distribution and microscopic images of PPI-stabilized emulsion and PPM-stabilized Pickering emulsion during in vitro digestion were measured according to the methods in Sections 2.3.1 and 2.5.2, respectively.

Free Fatty Acids (FFA) Release of Emulsions

During simulated intestinal digestion, lipids were continuously hydrolyzed to *FFA* by pancreatic lipase. The amount of FFAs released from the system was calculated based on the volume of the NaOH solution (0.1 mol/L) consumed in 2 h according to Equation (7), written as

$$FFA\% = 100 \times \frac{V_{NaOH} \times 0.1 \times M_{Lipid}}{2 \times m_{Lipid}} \tag{7}$$

where $M_{Lipid}$ is the molar mass of corn oil (872 g/mol) and $m_{Lipid}$ is the weight of corn oil in the emulsions (g).

## 2.6. Ice Cream Preparation and Characterization

### 2.6.1. Ice Cream Preparation

Different substitution rates (0%, 20%, 40%, 60%, 80%, and 100%) of cream in the ice cream were substituted by PPM(pH 7.0)-stabilized emulsion (Table 1). And the samples were named PR0, PR20, PR40, PR60, PR80, and PR100. They were homogenized for 5 min at 3000 rpm after mixing and then were pasteurized at 72 °C for 30 min. They were cooled and stored at 4 °C for 4 h. Finally, the finished product was obtained after pouring them into an ice cream maker (Pink Bunny, Zhongshan, China). Three parallel experiments for each sample were carried out.

**Table 1.** Composition of ice cream at different saturated fat replacer ratios.

| Material | Samples | | | | | |
|---|---|---|---|---|---|---|
| | **PR0** | **PR20** | **PR40** | **PR60** | **PR80** | **PR100** |
| Water/mL | 225 | 225 | 225 | 225 | 225 | 225 |
| Skim milk/g | 36 | 36 | 36 | 36 | 36 | 36 |
| Cream/g | 27 | 21.6 | 16.2 | 10.8 | 5.4 | 0 |
| Emulsion/g | 0 | 5.4 | 10.8 | 16.2 | 21.6 | 27 |
| Sucralose/g | 0.04 | 0.04 | 0.04 | 0.04 | 0.04 | 0.04 |
| Sodium alginate/g | 0.9 | 0.9 | 0.9 | 0.9 | 0.9 | 0.9 |

2.6.2. Rheological Properties

Rheological properties of ice cream slurries were measured according to the method of Section 2.5.3.

2.6.3. Melting Measurement

After extrusion of ice cream, about 30 g of ice cream (the mass was recorded as $M_0$) was weighed and placed on a metal net with a sieve size of 3 mm × 3 mm at 25 °C [28]. The dropping mass of ice cream was weighed every 10 min until it reached 50 min. Melting and melting rate were calculated according to Equations (8) and (9), written as

$$\text{Melting } (\%) = \frac{M_0}{M_{50}} \times 100\% \tag{8}$$

$$\text{Melting rate } (\text{g/min}) = \frac{\text{Mass of dropping}}{\text{Melting time}} \tag{9}$$

where $M_{50}$ is the mass of ice cream after 50 min of dropping.

2.6.4. Overrun Measurement

The overrun of ice cream was measured by recording per volume weight of ice cream slurry (m) and per volume weight of ice cream ($m_f$) according to Equation (10) [29].

$$\text{Overrun}(\%) = \frac{m - m_f}{m_f} \times 100\% \tag{10}$$

2.6.5. Sensory Evaluation

The sensory evaluation methods were based on the methods of Roberta et al. [30], with some modifications. According to the Chinese standard [31] (criterion for sensory evaluation of frozen drinks and ice cream), 20 reviewers were trained for 20 days to form the review panel. Samples were divided as separate portions in 30 mL plastic cups and were assessed in duplicate. The samples were coded, and their order randomized. Ice cream samples were prepared and stored at −18 °C. Before evaluation, the samples were left at room temperature for 5 min. The samples were evaluated by the reviewers in turn for the following five main aspects: color, morphology, organization, taste and odor, and impurity. Distilled water was required to eliminate the effects between samples after different flavors. A 100-point intensity scale (1–10 points for color, 6–30 points for morphology, 5–30 points for organization, 5–20 points for taste and odor, and 3–10 points for impurity) was used for each term, as detailed in Table 2. Sensory acceptance was obtained by adding up the scores of the color, morphology, organization, taste and odor, and impurity.

**Table 2.** Sensory evaluation scoring index and criteria of ice cream.

| Item | Sensory Evaluation Criteria | Scoring Criteria |
|---|---|---|
| Color | White or light yellow and uniform. | 9~10 |
| | Color is too light or too dark. | 5~8 |
| | There are different colors. | 1~4 |
| Morphology | Complete shape, uniform volume, no distortion. | 27~30 |
| | Incomplete shape, general uniform volume, slightly twisted and deformed. | 20~26 |
| | Appearance of clots or too sticky and other undesirable phenomena. | 11~19 |
| | Distorted appearance, incomplete, uneven. | 6~10 |

Table 2. *Cont.*

| Item | Sensory Evaluation Criteria | Scoring Criteria |
|---|---|---|
| Organization | Soft and dense, no air holes appear. | 28~30 |
| | Ice crystals are found in the tissue. | 21~27 |
| | Serious ice crystals are found in the tissue. | 14~20 |
| | The tissue structure is not fluffy enough or too fluffy. | 5~13 |
| Taste and odor | Soft milky flavor, no bean smell. | 17~20 |
| | Insufficient light milky flavor, with soya odor or too sweet. | 13~16 |
| | Bean-like flavor is serious or other foreign odor appears. | 5~12 |
| Impurity | No visible impurities. | 7~10 |
| | There are visible impurities. | 3~6 |

### 2.7. Statistical Analysis

All the experiments were repeated at least three times. All data were assed with one-way ANOVA followed with Tukey's multiple-range test. Differences were considered statistically significant at $p < 0.05$. The results were resented as mean ± standard deviations.

## 3. Results and Discussion

### 3.1. Protein Structure Analysis

#### 3.1.1. Particle Size Distribution of PPMs

The mean particle sizes of the PPMs prepared at different pH ranging from 4.0 to 7.0 are shown in Figure 1A. It shows that the pH has a significant effect on the particle size distribution of the prepared PPMs. During microparticulation, pea protein was thermally denatured to form a gel structure. After high-speed homogenization, the gel was crushed into a three-dimensional mesh structure with a diameter of 13.65–56.78 μm. The mean particle size of PPM (pH 6.0) was the maximum and was 56.78 μm. A decline of mean particle size was observed when the microparticulation pH was far from pH 6.0. Mean particle sizes of PPM (pH 4.0), PPM (pH 4.5), and PPM (pH 5.0) were 20.69 μm, 18.79 μm, and 20.80 μm, respectively. There was no significant difference among the mean particle sizes of PPM (pH 4.0), PPM (pH 4.5), and PPM (pH 5.0). The sample with the smallest particle size was PPM (pH 7.0), with a mean particle size of 13.65 μm.

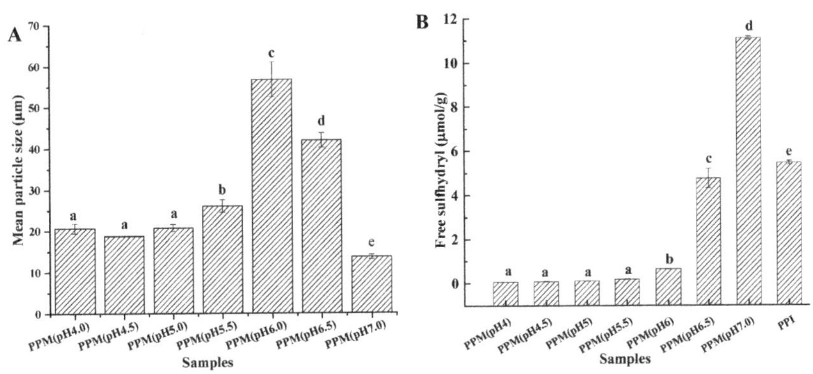


Figure 1. *Cont.*

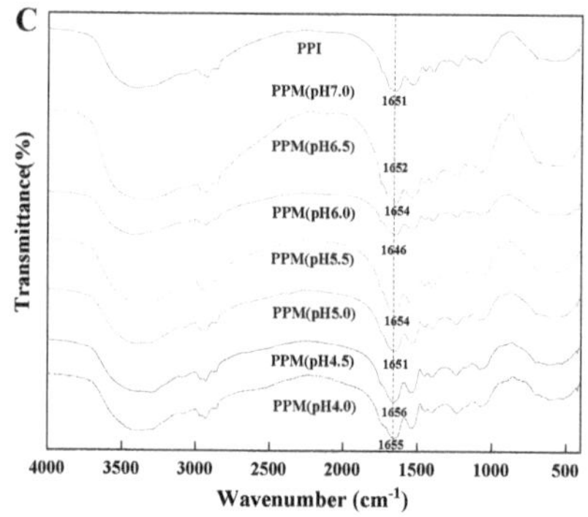


**Figure 1.** Particle size distribution (**A**), free sulfhydryl content (**B**) and FTIR spectrum (**C**) of pea protein micro−gels (PPMs) and pea protein isolate (PPI). The values that do not bear the same letter on the column are significantly different ($p < 0.05$).

Guo et al. [32], Cui et al. [33], and Schmitt et al. [34] confirmed that the surface charge is close to zero at the isoelectric point (pI). Thus, protein molecules aggregate and precipitate under hydrophobic interactions. However, when the pH is far away from the pI, the surface charge of protein molecules increases, which generates intermolecular repulsion and inhibits aggregation, leading to a decrease in protein particle size.

### 3.1.2. Content of Free Sulfhydryl Groups

Free sulfhydryl groups ($SH_F$) reflect the changes of the protein tertiary structure and played key role in the protein rigid structure formation. Free sulfhydryl groups can form disulfide bonds after oxidation, and disulfide bonds can be further reduced to free sulfhydryl groups through hydrogenation under the conditions of mutual conversion during heat treatment [35] and shear treatment.

The free sulfhydryl group content of different PPMs is shown in Figure 1B. $SH_F$ content of PPI is 5.45 μmol/g. The $SH_F$ content of PPM (pH 4.0), PPM (pH 4.5), PPM (pH 5.0), PPM (pH 5.5), PPM (pH 6.0), and PPM (pH 6.5) were 0.07, 0.09, 0.11, 0.19, 0.64, and 4.75 μmol/g, which were significantly lower than PPI; it was due to the highly concentrated, accessible, and reactive free sulfhydryl groups that were oxidized to form the disulfide bond [36].

The $SH_F$ content of PPMs gradually increased with the increasing microparticulation pH. The $SH_F$ content of PPM (pH 7.0) was the highest (11.10 μmol/g) and significantly higher than PPI. The increasing content of the free sulfhydryl groups may be due to the unfolding of the protein molecular structure and the break of disulfide bonds caused by high-temperature heating. Mession et al. [37] confirmed that heating causes the unfolding of protein molecules and exposing of embedded free sulfhydryl groups, disulfide bonds, and hydrophobic groups. The change of the functional group distribution can enhance the intermolecular interaction and then promote the formation of the gel network [38].

Guo et al. [10] found that the $SH_F$ content of protein molecules increases the promotion of the interaction between protein molecules and forms a tight and uniform network of gels. In summary, pH and heat can change the content of sulfhydryl bonds and affect the functional properties of proteins.

### 3.1.3. Fourier Transform Infrared Spectroscopy Analysis of PPI and PPMs

Figure 1C displays the spectra of PPI and PPMs in the 4000 to 400 $cm^{-1}$ range. The OMNIC software was used to fit the peaks in the range of 1700 to 1600 $cm^{-1}$ to calculate the content of each secondary structure of proteins [39]. The peaks located at 1650–1658 $cm^{-1}$ reflect the α-helix, 1610–1640 $cm^{-1}$ reflect the β-sheet, 1660–1700 $cm^{-1}$ reflect the β-turn, and the peaks at 1640–1650 $cm^{-1}$ reflect the random coil [40]; the results are shown in Table 3. As can be seen in Table 3, the PPI contains 11.66% α-helix, 33.04% β-sheet, 46.94%

β-turn, and 8.36% random coil. The β-sheet and β-turn are the dominant secondary structures of PPI and PPMs, followed by random coil and α-helix. Compared with PPI, the proportion of α-helix significantly decreased, which is consistent with the research of Beck et al. [41]. The β-sheet content in PPM (pH 4.0) and PPM (pH 4.5) significantly increased to 41.64% and 39.73%, respectively; the β-turn content had significantly decreased to 39.53% and 41.76%, respectively; random coil content showed no significant difference with PPI. The decreasing α-helix contents and increasing random coil content and β-sheet content were observed in PPM (pH 5.0) and PPM (pH 6.0). The secondary structures of PPM (pH 5.5) and PPM (pH 6.5) shifted from β-sheet and α-helix to β-turn and random coil, which is consistent with the results of Sun et al. [16]. It can be concluded that the reduction in α-helical structures might have been due to the fact that microparticulation destroyed the rigid structure of protein molecules, causing unfolding and rearrangement. The rearranged protein molecules tend to have a higher degree of intermolecular flexibility [39]. There is no significant difference in the content of each secondary structure between PPM (pH 7.0) and PPI. Therefore, it can be concluded that the different pH during macroparticulation influenced the degree of protein unfolding, the distribution of amino acid side chains, and changes in protein surface charges and [42]. Thus, flexible structures in the protein micro-gel particle formed after the disruption of the natural structures during microparticulation could lead to differences in their physiochemical properties with PPI, such as interfacial properties [16].

**Table 3.** Content of secondary structures of PPI and PPMs.

| Samples | α-Helix | β-Sheet | β-Turn | Random Coil |
|---|---|---|---|---|
| PPI | $11.66 \pm 0.20^{a}$ | $33.04 \pm 0.30^{cg}$ | $46.94 \pm 0.31^{c}$ | $8.36 \pm 0.20^{b}$ |
| PPM (pH 4.0) | $10.06 \pm 0.26^{b}$ | $41.64 \pm 0.55^{a}$ | $39.53 \pm 1.49^{e}$ | $8.77 \pm 0.67^{b}$ |
| PPM (pH 4.5) | $9.84 \pm 0.63^{b}$ | $39.73 \pm 0.09^{b}$ | $41.76 \pm 1.44^{d}$ | $9.04 \pm 0.21^{b}$ |
| PPM (pH 5.0) | $9.19 \pm 0.40^{b}$ | $32.29 \pm 0.33^{c}$ | $48.35 \pm 0.20^{b}$ | $10.16 \pm 0.13^{a}$ |
| PPM (pH 5.5) | $10.66 \pm 0.76^{b}$ | $28.57 \pm 0.96^{d}$ | $51.83 \pm 0.40^{a}$ | $8.94 \pm 0.60^{b}$ |
| PPM (pH 6.0) | $9.63 \pm 0.07^{b}$ | $35.47 \pm 1.24^{e}$ | $46.01 \pm 1.50^{c}$ | $8.89 \pm 0.19^{b}$ |
| PPM (pH 6.5) | $8.78 \pm 0.34^{c}$ | $30.40 \pm 0.09^{f}$ | $50.01 \pm 0.13^{a}$ | $10.81 \pm 0.56^{a}$ |
| PPM (pH 7.0) | $10.93 \pm 0.11^{ab}$ | $32.79 \pm 0.15^{cg}$ | $47.63 \pm 0.24^{bc}$ | $8.66 \pm 0.20^{b}$ |

The values that do not bear the same letter in the same column are significantly different ($p < 0.05$).

### 3.2. Emulsifying Properties

Figure 2A shows that the EAI of PPI increased gradually with the increase of pH from 5.5 to 7.0. Different pH preparation of PPMs also had a significant effect on EAI (Figure 2A). The EAI at pH near the pI (pH 4.0–6.0) was lower than the EAI at pH 6.0–7.0. The EAI values of the protein micro-gel particles prepared at pH 6.0–7.0 were higher than that of pI (pH 4.0–6.0), which was mainly due to the higher degree of protein molecules unfolding. The unfolded protein molecules could rapidly adsorb on the oil–water interface to form a denser viscoelastic interfacial layer, which stabilized the dispersed oil droplets. At neutral pH far from the isoelectric point, pea protein forms a stronger and denser viscoelastic network when adsorbed at the oil–water interface than that at pI (pH 4.0–6.0) [43]. Near the isoelectric point, protein molecules have no surface charge, so low intermolecular repulsion leads to molecular aggregation, which caused the lower EAI of proteins [44]. Another reason could be the proper three-phase contact angle of the micro-gel particle, which is attributed to balanced distribution of hydrophilic and hydrophobic groups of PPM prepared at pH 6.0–7.0 [33].

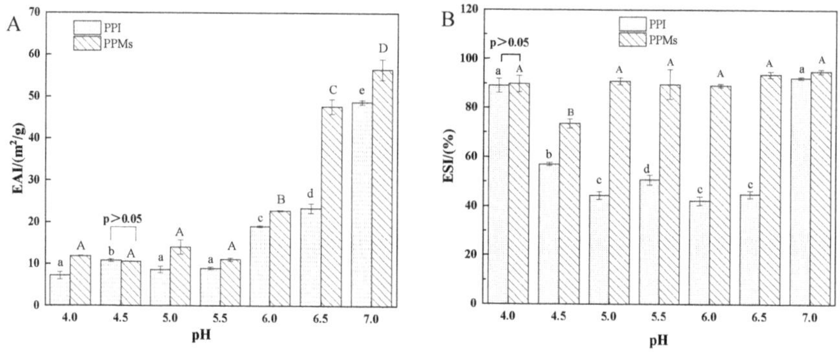


**Figure 2.** Emulsifying activity index (**A**) and emulsion stability index (**B**) of PPI at different pH and PPMs. The values that do not bear the same letter on the column are significantly different ($p < 0.05$).

The pH values of the dispersions have significant impacts on the ESI of PPI (Figure 2B). Among them, PPI has the better ESI, at pH 4.0 and pH 7.0, than others, with values of 89.10% and 92.40%, respectively. Except for PPM (pH 4.0), different preparation pH had no significant effect on the ESI values of other PPM samples. It can be seen that protein micro-gel particles once adsorbed on the oil–water interface, form a stable interfacial layer to stabilize the oil droplets [45].

The EAI and ESI of PPMs were both significantly higher than that of the PPI at corresponding pH (Figure 2), indicating that micro-gel particles have higher potential applications in multiphase systems than PPI. Natural pea proteins are globular proteins with well-defined molecular conformations, where hydrophilic groups are distributed on the surface of the protein molecule, while the hydrophobic groups are embedded within the molecule. During microparticulation, protein molecules partially unfolded, then recombined and cross-linked to form the gel network structure [36]. Subsequently, high-speed shear treatment disrupts the gel maintained by non-covalent bonds to obtain micro-gel particles. Simultaneously, the surface hydrophilic and hydrophobic groups of protein molecules undergo rearrangement [45]. In summary, compared to the rigid structure of PPI, more flexible structures were formed during microparticulation, which can be better adsorbed on the oil–water interface to form an interfacial layer that stabilizes the oil droplets. The higher EAI (56.56 $m^2/g$) and ESI (95.10%) of PPM (pH 7.0) than other samples (Figure 2) indicated that PPM (pH 7.0) is more suitable for application in O/W emulsion stabilization. Therefore, PPM (pH 7.0) was chosen for subsequent emulsion preparation and research.

### 3.3. *Characterization of the PPM (pH 7.0)-Stabilized Pickering Emulsion*

#### 3.3.1. Micromorphology Observation

After staining with Nile red, the oil droplets showed red after excitation at 636 nm. It can be seen from Figure 3A that oil droplets exist in the emulsion in a round shape with uniform size. There is an interfacial protein layer, which can not only effectively reduce the surface tension, but also help to form uniform and stable emulsion [46]. This also explains well the better ESI of PPM (pH 7.0). And it can be found that the droplets size of the emulsion was less than 10 μm, which indicated that the PPM (pH 7.0)-stabilized emulsion can simulate the smooth and delicate texture of saturated fat and can be used as a saturated fat replacement to replace thin cream to prepare low-saturated fat ice cream.

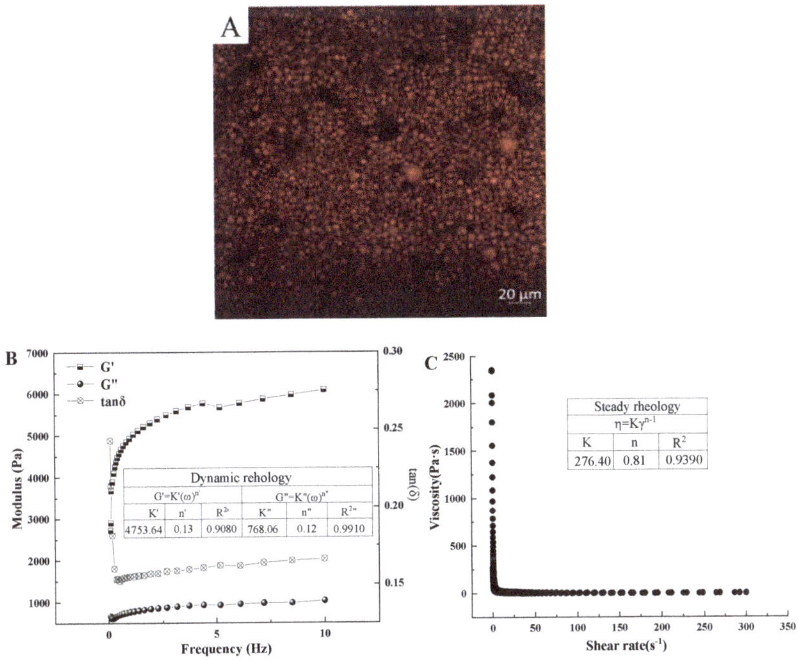


**Figure 3.** Microstructure (**A**), dynamic mechanical spectra, (**B**) and steady flow curves (**C**) of the PPM (pH 7.0)-stabilized Pickering emulsion.

3.3.2. Rheological Properties

Dynamic rheological measurements can reflect the formation and destruction of composite gel matrix [24]. The viscoelasticity of the PPM (pH 7.0)-stabilized Pickering emulsion can be characterized by storage modulus ($G'$) and loss modulus ($G''$). Figure 3B shows that $G'$ and $G''$ of the PPM (pH 7.0)-stabilized Pickering emulsion slightly increased with the increasing frequency. The slopes of $G'$ ($K' = 4753.64$ Pa·$s^{n'}$) were higher than that of $G''$ ($K'' = 768.06$ Pa·$s^{n''}$), which indicated that $G'$ was more frequency dependent than $G''$ [10]. $G'$ of the PPM (pH 7.0)-stabilized Pickering emulsion was always larger than $G''$ and tan $\delta$ was less than one throughout the tested frequency range, indicating that elastic behavior dominated the rheological properties [47].

Figure 3C shows that the viscosity of the PPM (pH 7.0)-stabilized Pickering emulsion decreased with the increase in shear rate, indicating that the emulsion is a shear-thinning fluid. In PPM (pH 7.0)-stabilized Pickering emulsion, the droplets are close enough together to interact with each other which may lead to the formation of a three-dimensional network of aggregated droplets. As the shear rate is increased, the hydrodynamic forces cause aggregates to become deformed and eventually disrupted which results in a reduction in the viscosity. The flow behavior index $n$ is 0.81, which is less than 1. It indicated that the PPM (pH 7.0)-stabilized Pickering emulsion is pseudoplastic fluid. According to Long et al. [48], cream is a pseudoplastic fluid and exhibits shear-thinning behavior. PPM (pH 7.0)-stabilized Pickering emulsion showed the same rheology properties as cream, which ensured its potential application as a saturated fat replacement for cream.

3.3.3. In Vitro Digestion

Figure 4A,B shows the particle size distribution and microstructure of PPI-stabilized emulsion and PPM (pH 7.0)-stabilized Pickering emulsion during simulated in vitro digestion. Figure 4$A_1$,$B_1$ shows that before digestion, the oil droplets in both emulsions exhibited uniformly distribution in the form of round balls. The particle sizes $d_{32}$ in the two emulsions are not significantly different. However, there is a very small distribution peak at 10–100 μm in the particle size distribution curve of the PPI-stabilized emulsion, indicating the presence of a few large oil droplets, which leads to a significantly higher $d_{43}$ (8.52 μm) than that of the PPM (pH 7.0)-stabilized Pickering emulsion (4.45 μm). Comparing Figure 4$A_2$,$B_2$ and Figure 4$A_1$,$B_1$, it can be found that after a short period of oral digestion, the particle size distribution and microscopic morphology of the two emulsions did not change significantly ($p > 0.05$). This can also be confirmed by the non-significantly different values of $d_{32}$ and $d_{43}$.

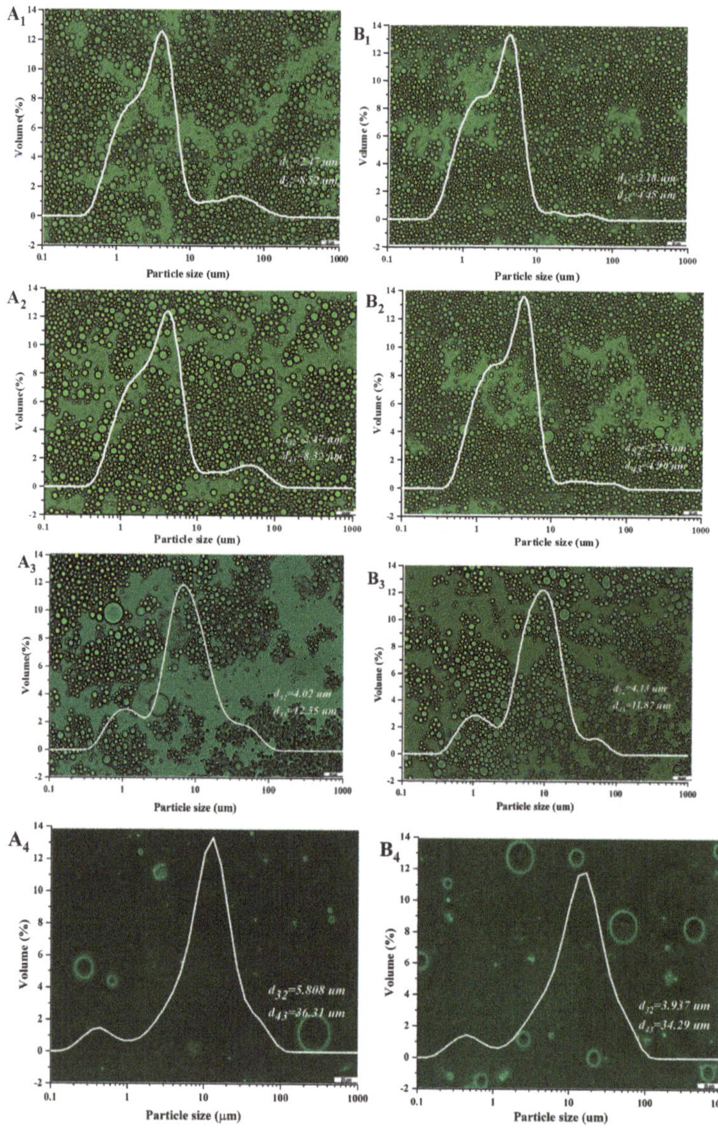


**Figure 4.** *Cont.*

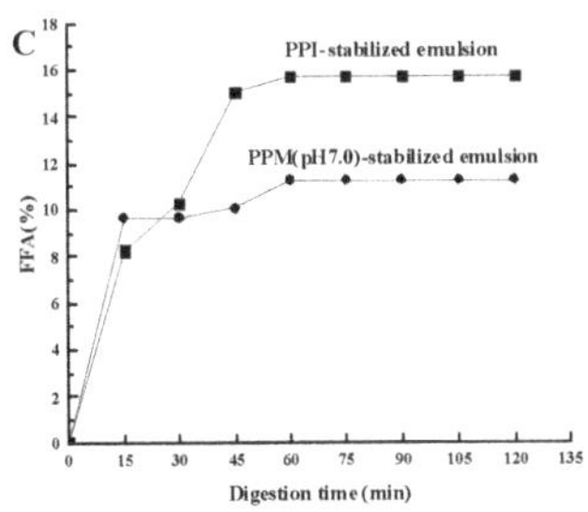


**Figure 4.** The microstructure, particle size distribution, and free fatty acids (FFA) release (**C**) of the PPI−stabilized emulsion (**A**) and PPM (pH 7.0)-stabilized Pickering emulsion (**B**) during in vitro digestion. (**A$_1$,B$_1$**) Initial, (**A$_2$,B$_2$**) Oral phase, (**A$_3$,B$_3$**) Gastric phase, (**A$_4$,B$_4$**) Intestinal phase.

After simulated gastric digestion, the particle size distribution and microscopic morphology of both emulsions changed dramatically (Figure 4A$_3$,B$_3$). The oil droplet size increased significantly ($p < 0.05$), which was attributed to the coalescence of the oil droplets. The gastric acid (pH 2.0) and pepsin in the SGF caused the protein at the oil–water interface in the emulsions to be denatured, losing the role of forming interfacial film to protect the oil droplets. It further led to flocculation and even coalescence of the oil droplets [49]. Compared to PPM (pH 7.0)-stabilized Pickering emulsions, PPI-stabilized emulsions showed more flocculation of oil droplets after simulated gastric digestion (Figure 4A$_3$,B$_3$). This suggests that the micro-gel particles formed by pea proteins have an enhanced ability to resist both the acidic environment and pepsin digestion [50].

After simulated intestinal digestion, the number of oil drops in the two emulsions decreased significantly under the dilution of digestive fluid and the digestion of lipase-colipase (Figure 4A$_4$,B$_4$); and the oil droplet size significantly increased ($p < 0.05$). However, the oil droplet size in PPM (pH 7.0)-stabilized Pickering emulsion was significantly smaller than that in PPI-stabilized emulsion ($p < 0.05$). And the number of oil droplets was significantly greater than that in PPI stabilized emulsion ($p < 0.05$). It indicated that PPM (pH 7.0) has the ability to protect the integrity of the emulsion structure in the simulated gastrointestinal fluid and resist the digestion of simulated gastrointestinal fluid. This can also be confirmed from Figure 4C; it shows that the FFA release rate of both emulsions increases rapidly within 45 min of simulated intestinal digestion. After 45 min, the FFA release rate slowed down and eventually approached 0. Obviously, the FFA release rate of PPM (pH 7.0)-stabilized Pickering emulsion was significantly less than that of the PPI-stabilized emulsion ($p < 0.05$). This indicates that the PPM (pH 7.0) on the oil–water interface has the ability to block the adsorption of bile salts and lipase, thus the delay of the lipid digestion of emulsions. The results of Sarkar et al. [51] also highlighted that the protein gel is able to slow the diffusion of both lipase–colipase and proteases, thus inhibiting lipid digestion of O/W emulsion. Therefore, PPM (pH 7.0)-stabilized emulsion can be used as compound fat replacement to reduce the lipid content of food and decline the lipid bioavailability.

### *3.4. Steady-State Shear Properties of Ice Cream Slurry*

As can be seen from Figure 5, the viscosity of all ice cream slurries show a decreasing trend with the increasing shear rate and show a nonlinear relationship, indicating that all ice cream slurries exhibited shear-thinning behavior. This is due to the fact that at rest or low flow rates, the macromolecular chain-like structures are entangled with each other, and the viscosity of all ice cream slurries is high. With the increasing shear rate, due to the effect of the shear stress, the disordered macromolecular chain rolled and rotated and contracted into clusters, which reduced the intermolecular cross-linking, and thus the viscosity gradually decreased [52].

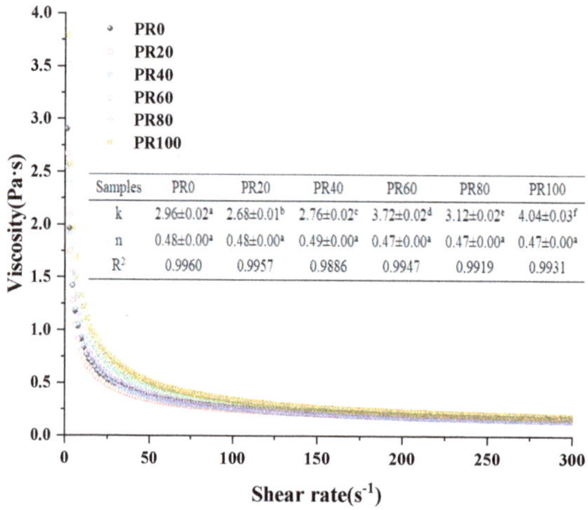


| Samples | PR0 | PR20 | PR40 | PR60 | PR80 | PR100 |
|---|---|---|---|---|---|---|
| k | 2.96±0.02[a] | 2.68±0.01[b] | 2.76±0.02[c] | 3.72±0.02[d] | 3.12±0.02[e] | 4.04±0.03[f] |
| n | 0.48±0.00[a] | 0.48±0.00[a] | 0.49±0.00[a] | 0.47±0.00[a] | 0.47±0.00[a] | 0.47±0.00[a] |
| $R^2$ | 0.9960 | 0.9957 | 0.9886 | 0.9947 | 0.9919 | 0.9931 |

**Figure 5.** Flow curves of ice cream slurries measured at 25 °C. [a–f]: The values that do not bear the same letter on the column are significantly different ($p < 0.05$).

According to the power law model with high regression coefficients ($R^2$ = 0.98–0.99), the fitting parameters were used to obtain the flow behavior index (*n*) and consistency coefficient (K) of all ice cream slurries. As can be seen in Figure 5, there were no significant differences in the n values of the ice cream slurries as the substitution rate increased from 0% to 100%. The index n is greater than zero and less than one, indicating that all ice cream slurries are pseudoplastic fluids. K values significantly increased with the increasing substitution rate, which indicated that the ice cream slurries showed more viscous properties due to the increase in apparent viscosity. Wang et al. [53] also found that the K values of ice cream slurries increased with the increase of soybean oil body substitution rate. They claimed that it was due to the fact that the increased content of Pickering emulsion introduces PPM biomolecules, whose hydrogen bonding or hydrophobic interactions between the nonpolar side segments of the carbon skeleton produce new ordered and improved structures, which lead to differences in rheological behavior.

### 3.5. Dynamic Oscillatory Measurements of Ice Cream Slurry

The dynamic oscillatory of ice cream slurries can reflect the physicochemical properties and flow behavior during chewing in the mouth. Storage modulus $G'$ and loss modulus $G''$ were analyzed, shown in Figure 6A,B. $G'$ and $G''$ of all ice cream slurries, except PR40 and PR60, increased with increasing frequency until 1.3 Hz, which indicated the destruction of the structures. $G'$ and $G''$ of PR40 and PR60 remained stable until 2.8 Hz, indicating that the structure was still intact. This also indicates the enhanced structural stability of low-fat ice cream slurries at 40% and 60% substitution rates. The tan δ ($G''/G'$) of ice cream slurries is a key indicator of the phase melting and structural changes [54], which are shown in Figure 6C. The tan δ of all the samples in the measured frequency range are less than one, indicating that the $G'$ values are higher than the $G''$ values, which suggests that all the ice cream slurries dominate the liquid characteristics, which are also known as viscoelastic fluids; viscoelastic fluid characteristics are the desired rheological properties of ice cream. It can be seen that the application of PPM (pH 7.0)-stabilized Pickering emulsion replacement of saturated fat in ice cream did not change its solid/liquid behavior. This also indicates PPM (pH 7.0)-stabilized Pickering emulsion has the potential to be applied in low-fat foods as compound fat replacements.

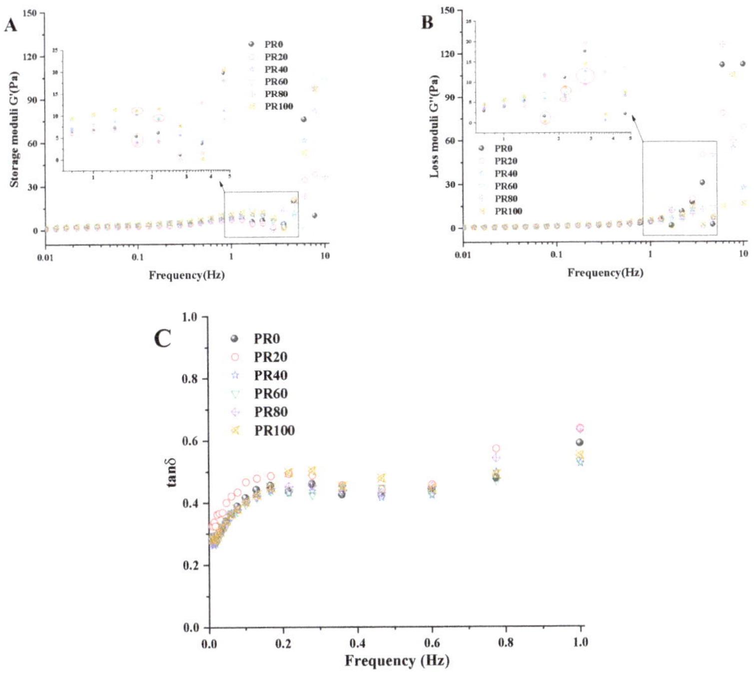


**Figure 6.** Dynamic mechanical spectra ((**A**) storage modulus, (**B**) loss modulus, (**C**) tan δ) of ice cream slurries.

## 3.6. Physicochemical Properties and Sensory Evaluation of Ice Cream

Table 4 shows the effects of the substitution rates of the PPM (pH 7.0)-stabilized Pickering emulsion on the physicochemical properties and sensory evaluation of the ice cream. There was no significant difference in the melting of each ice cream as the substitution rate increased to 80%. This indicates that the substitution of saturated fat by a moderate amount of PPM (pH 7.0)-stabilized Pickering emulsion does not affect the ice phase organization. However, the melting of PR100 (1.345%) was significantly greater than the other samples, due to the fact that the ice phase arrangement of PR100 was different from the other samples. In the PR100 with a lower overrun and stability, there was not enough network space structure to prevent melting and complete collapse. According to Kaleda et al. [55], the difference in melting between the two types of ice cream can only be attributed to differences in ice phase arrangement.

**Table 4.** Effects of fat substitution ratio on physicochemical indexes and sensory evaluation of ice cream.

| Samples | Melting (%) | Melting Rate (g/min) | Overrun (%) | Sensory Evaluation |
|---|---|---|---|---|
| PR0 | $0.64 \pm 0.02^{a}$ | $0.23 \pm 0.00^{a}$ | $85.70 \pm 0.71^{a}$ | $93.7 \pm 1.2^{a}$ |
| PR20 | $0.63 \pm 0.02^{a}$ | $0.24 \pm 0.02^{a}$ | $131.90 \pm 3.25^{b}$ | $89.7 \pm 0.6^{b}$ |
| PR40 | $0.78 \pm 0.05^{a}$ | $0.25 \pm 0.01^{a}$ | $103.70 \pm 9.62^{c}$ | $92.0 \pm 1.0^{ac}$ |
| PR60 | $0.50 \pm 0.02^{a}$ | $0.31 \pm 0.01^{b}$ | $93.13 \pm 1.41^{acd}$ | $95.0 \pm 0^{a}$ |
| PR80 | $0.40 \pm 0.01^{a}$ | $0.46 \pm 0.04^{c}$ | $82.70 \pm 2.12^{ade}$ | $83.0 \pm 1.0^{e}$ |
| PR100 | $1.35 \pm 0.40^{f}$ | $0.33 \pm 0.02^{b}$ | $94.75 \pm 2.90^{acd}$ | $91.8 \pm 1.5^{c}$ |

The values do not bear the same letter in the same column are significantly different ($p < 0.05$).

The melting rate results showed that there was no significant difference among the melting rates of PR20, PR40, and PR0. However, as the substitution rate continued to increase to 80%, the melting rate significantly increased to 0.46 g/min. This may be related to the fact that protein micro-gel particles in the emulsion increase the protein content of the ice cream, which alters the melting rate of the final product [56]. When the PPM (pH 7.0)-stabilized Pickering emulsion completely replaced the saturated fat in the ice cream, the melting rate was 0.33 g/min rather lower than PR80.

With an increase in the substitution ratio, the overrun of PR20 and PR40 increased from 85.70% of PR0 to 131.90% of PR20 and 103.70% of PR40, respectively. The overrun of PR20 and PR40, respectively, was 131.90% and 103.70%, which was significantly higher than the other samples. The higher overrun of PR20 and PR40 was related to the lower viscosity than other samples (Figure 5). The overrun was positively correlated with the amount of air introduced during the ice cream production [57]. The low viscosity of PR20 and PR40 facilitated a large amount of air during vigorous mixing. The ice cream with suitable overrun (90–110%) had a soft and delicate taste, a strong melting resistance, an unstable shape, and a difficult deformation [58]. The overrun of PR40, PR60, and PR100 were all within this range, which may result in higher sensory evaluation than PR20 and PR80 (Table 4).

The presence of a slight beany smell was the main reason why the sensory evaluation of PR100 was significantly lower than that of PR40 and PR60. The obvious ice crystal taste was the main reason for the lowest sensory score of PR80. The PR0 had a higher sensory evaluation and no significant difference compared to PR40 and PR60, which was due to its higher saturated fat content. Mahdian and Karazhian [59] confirmed that the sensory evaluation of ice cream was related to the content of cream. Reducing saturated fat of ice cream could result in structural defects, such as ice crystals and a rough texture.

## 4. Conclusions

PPM (pH 7.0) was suitable for application in O/W emulsion stabilization because of its proper particle size, more flexible structure, and high EAI and ESI. The Pickering emulsion stabilized by PPM (pH 7.0) had a uniform oil droplet distribution and similar rheological properties to cream; therefore, it can be used as a saturated fat replacement in the production of ice cream. PPM (pH 7.0)-stabilized emulsion can be used to substitute 60% cream to produce low-saturated fat ice cream, which had high structural stability, similar melting properties, overrun, and sensory properties with PR0. The results indicated that PPM (pH 7.0)-stabilized Pickering emulsion has a similar fatty acid profile to that of the corn oil used, but the rheological properties similar to that of saturated fatty acid. Therefore, it is feasible to use it as a saturated fat replacement in the production of low-saturated fat food to meet the demand of consumers for unsaturated fat.

**Supplementary Materials:** The following supporting information can be downloaded at: https://www.mdpi.com/article/10.3390/foods13101511/s1, Table S1: Stock solutions preparation of simulated digestion fluids.

**Author Contributions:** Conceptualization, supervision, and editing, C.S. and B.L.; writing—original draft and methodology, Y.G. and X.Q.; writing—review, X.Z.; validation, X.L.; formal analysis, C.J. All authors have read and agreed to the published version of the manuscript.

**Funding:** This work was funded by the National Natural Science Foundation of China (32302057, 32101904), the Natural Science Foundation of Shandong Province (ZR2020QC219, ZR2019BC036), and the 2023 Postgraduate Science and Technology Innovation Foundation of Yantai University (GGIFYTU2357).

**Institutional Review Board Statement:** The study was conducted in accordance with the Declaration of Helsinki, and approved by Yantai University Laboratory Animal Ethics Committee (protocol code 2024-090 and date of May 11, 2023).

**Informed Consent Statement:** Informed consent was obtained from all subjects involved in the study.

**Data Availability Statement:** The original contributions presented in the study are included in the article/Supplementary Material, further inquiries can be directed to the corresponding author.

**Conflicts of Interest:** The authors declare no conflicts of interest.

## References


1. Dagenais, G.R.; Leong, D.P.; Rangarajan, S.; Lanas, F.; Lopez-Jaramillo, P.; Gupta, R.; Diaz, R.; Avezum, A.; Oliveira, G.B.F.; Wielgosz, A.; et al. Variations in common diseases, hospital admissions, and deaths in middle-aged adults in 21 countries from five continents (PURE): A prospective cohort study. *Lancet* **2020**, *395*, 785–794. [CrossRef]
2. Lichtenstein, A.H. Dietary *Trans* Fatty Acids and Cardiovascular Disease Risk: Past and Present. *Curr. Atheroscler. Rep.* **2014**, *16*, 1–7. [CrossRef]
3. Torrejon, C.; Uauy, R. Quality of fat intake, atherosclerosis and coronary disease: Effects of saturated and *trans* fatty acids. *Rev. Medica De Chile* **2011**, *139*, 924–931. [CrossRef]
4. Briggs, M.A.; Petersen, K.S.; Kris-Etherton, P.M. Saturated Fatty Acids and Cardiovascular Disease: Replacements for Saturated Fat to Reduce Cardiovascular Risk. *Healthcare* **2017**, *5*, 29. [CrossRef]
5. Zetzl, A.K.; Marangoni, A.G.; Barbut, S. Mechanical properties of ethylcellulose oleogels and their potential for saturated fat reduction in frankfurters. *Food Funct.* **2012**, *3*, 327–337. [CrossRef]
6. Munk, M.B.; Munk, D.M.E.; Gustavsson, F.; Risbo, J. Using Ethylcellulose to Structure Oil Droplets in Ice Cream Made with High Oleic Sunflower Oil. *J. Food Sci.* **2018**, *83*, 2520–2526. [CrossRef]
7. Amador, J.; Hartel, R.; Rankin, S. The Effects of Fat Structures and Ice Cream Mix Viscosity on Physical and Sensory Properties of Ice Cream. *J. Food Sci.* **2017**, *82*, 1851–1860. [CrossRef]
8. Bourouis, I.; Pang, Z.; Liu, X. Recent advances on uses of protein and/or polysaccharide as fat replacers: Textural and tribological perspectives: A review. *J. Agric. Food Res.* **2023**, *11*, 100519. [CrossRef]
9. Zhang, M.; Sun, C.; Li, Q. Interaction Between the Polysaccharides and Proteins in Semisolid Food Systems. In *Encyclopedia of Food Chemistry*; Elsevier: Amsterdam, The Netherlands, 2019; pp. 439–445. [CrossRef]
10. Sun, C.; Liu, R.; Liang, B.; Wu, T.; Sui, W.; Zhang, M. Microparticulated whey protein-pectin complex: A texture-controllable gel for low-fat mayonnaise. *Food Res. Int.* **2018**, *108*, 151–160. [CrossRef]
11. Sun, C.; Liu, R.; Wu, T.; Liang, B.; Shi, C.; Cong, X.; Hou, T.; Zhang, M. Combined Superfine Grinding and Heat-Shearing Treatment for the Microparticulation of Whey Proteins. *Food Bioprocess Technol.* **2015**, *9*, 378–386. [CrossRef]
12. Paglarini, C.D.; Martini, S.; Pollonio, M.A.R. Using emulsion gels made with sonicated soy protein isolate dispersions to replace fat in frankfurters. *Lwt-Food Sci. Technol.* **2019**, *99*, 453–459. [CrossRef]
13. Lima, T.L.S.; da Costa, G.F.; Alves, R.D.; de Araujo, C.D.L.; da Silva, G.F.G.; Ribeiro, N.L.; de Figueiredo, C.F.V.; de Andrade, R.O. Vegetable oils in emulsified meat products: A new strategy to replace animal fat. *Food Sci. Technol.* **2022**, *42*, 103621. [CrossRef]
14. Lin, D.; Kelly, A.L.; Miao, S. Formation and creaming stability of alginate/micro-gel particle-induced gel-like emulsions stabilized by soy protein isolate. *Food Hydrocoll.* **2021**, *121*, 107040. [CrossRef]
15. Sun, C.; Liu, R.; Sheng, H.; Wang, R.; Zhang, Z.; Zhao, J.; Zhang, M. Effect of microparticulation and xanthan gum on the stability and lipid digestion of oil-in-water emulsions stabilized by whey protein. *Food Funct.* **2018**, *9*, 4683–4694. [CrossRef]
16. Sun, C.; Liang, B.; Sheng, H.; Wang, R.; Zhao, J.; Zhang, Z.; Zhang, M. Influence of initial protein structures and xanthan gum on the oxidative stability of O/W emulsions stabilized by whey protein. *Int. J. Biol. Macromol.* **2018**, *120*, 34–44. [CrossRef]
17. Klojdova, I.; Stathopoulos, C. The Potential Application of Pickering Multiple Emulsions in Food. *Foods* **2022**, *11*, 1558. [CrossRef]
18. Han, S.; Cui, F.Z.; McClements, D.J.; Xu, X.F.; Ma, C.C.; Wang, Y.T.; Liu, X.B.; Liu, F.G. Structural Characterization and Evaluation of Interfacial Properties of Pea Protein Isolate-EGCG Molecular Complexes. *Foods* **2022**, *11*, 2895. [CrossRef] [PubMed]
19. Morell, P.; Lopez-Garcia, A.; Hernando, I.; Quiles, A. Improving Pea Protein Emulsifying Capacity by Glycosylation to Prepare High-Internal-Phase Emulsions. *Foods* **2023**, *12*, 870. [CrossRef]
20. Yang, J.; Duan, Y.; Zhang, H.; Huang, F.; Wan, C.; Cheng, C.; Wang, L.; Peng, D.; Deng, Q. Ultrasound coupled with weak alkali cycling-induced exchange of free sulfhydryl-disulfide bond for remodeling interfacial flexibility of flaxseed protein isolates. *Food Hydrocoll.* **2023**, *140*, 108597. [CrossRef]
21. Hashemi, B.; Madadlou, A.; Salami, M. Functional and in vitro gastric digestibility of the whey protein hydrogel loaded with nanostructured lipid carriers and gelled via citric acid-mediated crosslinking. *Food Chem.* **2017**, *237*, 23–29. [CrossRef]
22. Figueroa-González, J.J.; Lobato-Calleros, C.; Vernon-Carter, E.J.; Aguirre-Mandujano, E.; Alvarez-Ramirez, J.; Martínez-Velasco, A. Modifying the structure, physicochemical properties, and foaming ability of amaranth protein by dual pH-shifting and ultrasound treatments. *LWT* **2022**, *153*, 112561. [CrossRef]
23. Yang, F.; Liu, X.; Ren, X.e.; Huang, Y.; Huang, C.; Zhang, K. Swirling cavitation improves the emulsifying properties of commercial soy protein isolate. *Ultrason. Sonochem.* **2018**, *42*, 471–481. [CrossRef]
24. Genccelep, H.; Saricaoglu, F.T.; Anil, M.; Agar, B.; Turhan, S. The effect of starch modification and concentration on steady-state and dynamic rheology of meat emulsions. *Food Hydrocoll.* **2015**, *48*, 135–148. [CrossRef]

25. Liang, H.-N.; Tang, C.-h. Pea protein exhibits a novel Pickering stabilization for oil-in-water emulsions at pH 3.0. *LWT-Food Sci. Technol.* **2014**, *58*, 463–469. [CrossRef]
26. Cheng, Y.; Donkor, P.O.; Yeboah, G.B.; Ayim, I.; Wu, J.; Ma, H.L. Modulating the in vitro digestion of heat-set whey protein emulsion gels via gelling properties modification with sequential ultrasound pretreatment. *LWT-Food Sci. Technol.* **2021**, *149*, 111856. [CrossRef]
27. Minekus, M.; Alminger, M.; Alvito, P.; Ballance, S.; Bohn, T.; Bourlieu, C.; Carriere, F.; Boutrou, R.; Corredig, M.; Dupont, D.; et al. A standardised static in vitro digestion method suitable for food—An international consensus. *Food Funct.* **2014**, *5*, 1113–1124. [CrossRef]
28. Tang, C.-H.; Liu, F. Cold, gel-like soy protein emulsions by microfluidization: Emulsion characteristics, rheological and microstructural properties, and gelling mechanism. *Food Hydrocoll.* **2013**, *30*, 61–72. [CrossRef]
29. Rolon, M.L.; Bakke, A.J.; Coupland, J.N.; Hayes, J.E.; Roberts, R.F. Effect of fat content on the physical properties and consumer acceptability of vanilla ice cream. *J. Dairy Sci.* **2017**, *100*, 5217–5227. [CrossRef]
30. Roberta, T.; Matteo, Z.; Giovanna, F.; Rodrigo, G.O.; Lucia, S.; Matteo, S.; Fabio, F. Dietary fibers effects on physical, thermal, and sensory properties of low-fat ice cream. *Food Sci. Technol.* **2024**, *199*, 0023–6438. [CrossRef]
31. *GB/T 31114-2014*; Frozen Drinks-Ice Cream. State General Administration of the People's Republic of China for Quality Supervision and Inspection and Quarantine Standardization Administration: Beijing, China, 2014.
32. Guo, J.; Zhou, Q.; Liu, Y.-C.; Yang, X.-Q.; Wang, J.-M.; Yin, S.-W.; Qi, J.-R. Preparation of soy protein-based microgel particles using a hydrogel homogenizing strategy and their interfacial properties. *Food Hydrocoll.* **2016**, *58*, 324–334. [CrossRef]
33. Cui, T.T.; Jia, A.R.; Yao, M.K.; Zhang, M.A.S.; Sun, C.C.; Shi, Y.P.; Liu, X.; Sun, J.M.; Liu, C.H. Characterization and Caco-2 Cell Transport Assay of Chito-Oligosaccharides Nano-Liposomes Based on Layer-by-Layer Coated. *Molecules* **2021**, *26*, 4144. [CrossRef] [PubMed]
34. Schmitt, C.; Bovay, C.; Rouvet, M. Bulk self-aggregation drives foam stabilization properties of whey protein microgels. *Food Hydrocoll.* **2014**, *42*, 139–148. [CrossRef]
35. Chen, X.; Guo, M.; Sang, Y.; Sun, J. Effect of ball-milling treatment on the structure, physicochemical properties and allergenicity of proteins from oyster (*Crassostrea gigas*). *LWT* **2022**, *166*, 113803. [CrossRef]
36. Peng, W.; Kong, X.; Chen, Y.; Zhang, C.; Yang, Y.; Hua, Y. Effects of heat treatment on the emulsifying properties of pea proteins. *Food Hydrocoll.* **2016**, *52*, 301–310. [CrossRef]
37. Mession, J.-L.; Roustel, S.; Saurel, R. Interactions in casein micelle—Pea protein system (part I): Heat-induced denaturation and aggregation. *Food Hydrocoll.* **2017**, *67*, 229–242. [CrossRef]
38. Wu, M.; Cao, Y.; Lei, S.; Liu, Y.; Wang, J.; Hu, J.; Li, Z.; Liu, R.; Ge, Q.; Yu, H. Protein structure and sulfhydryl group changes affected by protein gel properties: Process of thermal-induced gel formation of myofibrillar protein. *Int. J. Food Prop.* **2019**, *22*, 1834–1847. [CrossRef]
39. Wang, Z.; Zhao, H.; Tao, H.; Yu, B.; Cui, B.; Wang, Y. Ultrasound improves the physicochemical and foam properties of whey protein microgel. *Front. Nutr.* **2023**, *10*, 1140737. [CrossRef] [PubMed]
40. Wang, C.; Jiang, L.; Wei, D.; Li, Y.; Sui, X.; Wang, Z.; Li, D. Effect of Secondary Structure determined by FTIR Spectra on Surface Hydrophobicity of Soybean Protein Isolate. *Procedia Eng.* **2011**, *15*, 4819–4827. [CrossRef]
41. Beck, S.M.; Knoerzer, K.; Arcot, J. Effect of low moisture extrusion on a pea protein isolate's expansion, solubility, molecular weight distribution and secondary structure as determined by Fourier Transform Infrared Spectroscopy (FTIR). *J. Food Eng.* **2017**, *214*, 166–174. [CrossRef]
42. Nguyen, B.T.; Nicolai, T.; Chassenieux, C.; Schmitt, C.; Bovetto, L. Heat-induced gelation of mixtures of whey protein isolate and sodium caseinate between pH 5.8 and pH 6.6. *Food Hydrocoll.* **2016**, *61*, 433–441. [CrossRef]
43. Liang, H.-N.; Tang, C.-H. pH-dependent emulsifying properties of pea [*Pisum sativum* (L.)] proteins. *Food Hydrocoll.* **2013**, *33*, 309–319. [CrossRef]
44. Kornet, R.; Roozalipour, S.L.; Venema, P.; van der Goot, A.J.; Meinders, M.B.J.; van der Linden, E. Coacervation in pea protein solutions: The effect of pH, salt, and fractionation processing steps. *Food Hydrocoll.* **2022**, *125*, 107379. [CrossRef]
45. Han, K.; Feng, X.; Yang, Y.; Tang, X.; Gao, C. Changes in the physicochemical, structural and emulsifying properties of chicken myofibrillar protein via microfluidization. *Innov. Food Sci. Emerg. Technol.* **2023**, *83*, 103236. [CrossRef]
46. Wan, Z.-L.; Wang, L.-Y.; Wang, J.-M.; Zhou, Q.; Yuan, Y.; Yang, X.-Q. Synergistic interfacial properties of soy protein–stevioside mixtures: Relationship to emulsion stability. *Food Hydrocoll.* **2014**, *39*, 127–135. [CrossRef]
47. Cao, Y.; Li, Z.; Li, B.; Fan, X.; Liu, M.; Zhao, J. Mitigation of oxidation-induced loss of myofibrillar protein gelling potential by the combination of pyrophosphate and l-lysine. *LWT* **2022**, *157*, 113068. [CrossRef]
48. Long, Z.; Zhao, M.; Zhao, Q.; Yang, B.; Liu, L. Effect of homogenisation and storage time on surface and rheology properties of whipping cream. *Food Chem.* **2012**, *131*, 748–753. [CrossRef]
49. Zhang, X.R.; Guo, Y.X.; Liu, H.; Liang, B.; He, H.J.; Fu, X.J.; Sun, C.C.; Li, X.L.; Ji, C.J. Preparation, characterization of curdlan-based emulsion micro-gel particles and its application in low-fat pork sausages. *LWT-Food Sci. Technol.* **2023**, *185*, 115160. [CrossRef]
50. Deng, R.X.; Mars, M.; Van der Sman, R.G.M.; Smeets, P.A.M.; Janssen, A.E.M. The importance of swelling for in vitro gastric digestion of whey protein gels. *Food Chem.* **2020**, *330*, 127182. [CrossRef] [PubMed]
51. Sarkar, A.; Juan, J.M.; Kolodziejczyk, E.; Acquistapace, S.; Donato-Capel, L.; Wooster, T.J. Impact of Protein Gel Porosity on the Digestion of Lipid Emulsions. *J. Agric. Food Chem.* **2015**, *63*, 8829–8837. [CrossRef]

52. Akbari, M.; Eskandari, M.H.; Davoudi, Z. Application and functions of fat replacers in low-fat ice cream: A review. *Trends Food Sci. Technol.* **2019**, *86*, 34–40. [CrossRef]
53. Wang, W.; Wang, M.; Xu, C.; Liu, Z.J.; Gu, L.Y.; Ma, J.G.; Jiang, L.Z.; Jiang, Z.M.; Hou, J.C. Effects of Soybean Oil Body as a Milk Fat Substitute on Ice Cream: Physicochemical, Sensory and Digestive Properties. *Foods* **2022**, *11*, 1504. [CrossRef] [PubMed]
54. Freire, D.O.; Wu, B.Q.; Hartel, R.W. Effects of structural attributes on the rheological properties of ice cream and melted ice cream. *J. Food Sci.* **2020**, *85*, 3885–3898. [CrossRef] [PubMed]
55. Kaleda, A.; Tsanev, R.; Klesment, T.; Vilu, R.; Laos, K. Ice cream structure modification by ice-binding proteins. *Food Chem.* **2018**, *246*, 164–171. [CrossRef] [PubMed]
56. Moriano, M.E.; Alamprese, C. Organogels as novel ingredients for low saturated fat ice creams. *Lwt-Food Sci. Technol.* **2017**, *86*, 371–376. [CrossRef]
57. Kasapoglu, M.Z.; Sagdic, O.; Avci, E.; Tekin-Cakmak, Z.H.; Karasu, S.; Turker, R.S. The Potential Use of Cold-Pressed Coconut Oil By-Product as an Alternative Source in the Production of Plant-Based Drink and Plant-Based Low-Fat Ice Cream: The Rheological, Thermal, and Sensory Properties of Plant-Based Ice Cream. *Foods* **2023**, *12*, 650. [CrossRef] [PubMed]
58. Liu, R.; Wang, L.G.; Liu, Y.; Wu, T.; Zhang, M. Fabricating soy protein hydrolysate/xanthan gum as fat replacer in ice cream by combined enzymatic and heat-shearing treatment. *Food Hydrocoll.* **2018**, *81*, 39–47. [CrossRef]
59. Mahdian, E.; Karazhian, R. Effects of Fat Replacers and Stabilizers on Rheological, Physicochemical and Sensory Properties of Reduced-fat Ice Cream. *J. Agric. Sci. Technol.* **2013**, *15*, 1163–1174. Available online: http://jast.modares.ac.ir/article-23-4848-en.html (accessed on 1 November 2013).

foods

MDPI

Article

# Preparation of Edible Colorant Lake Using Calcium Carbonate and β-Carotene: Structural Characterization and Formation Mechanism Study

Yuhan Liu [1,2,3], Le Jing [1,2,3], Jiaqi Cui [1,2,3], Dongdong Yuan [1,2,3,*] and Chengtao Wang [1,2,3]

1 Beijing Engineering and Technology Research Center of Food Additives, School of Food and Health, Beijing Technology and Business University, Beijing 100048, China; liuyuhan_210@163.com (Y.L.); 2130022077@st.btbu.edu.cn (L.J.); c18647362147@163.com (J.C.); wctbtbu@hotmail.com (C.W.)

2 Key Laboratory of Geriatric Nutrition and Health, Ministry of Education, School of Food and Health, Beijing Technology and Business University, Beijing 100048, China

3 China Food Flavor and Nutrition Health Innovation Center, School of Food and Health, Beijing Technology and Business University, Beijing 100048, China

* Correspondence: dongdong.yuan@btbu.edu.cn

**Abstract:** This study prepared a novel β-carotene colorant lake using calcium carbonate ($CaCO_3$) and investigated the lake formation process and its basic characteristics. Kinetic adsorption analysis confirmed that medium pH (9.3) and medium temperature (40 °C) were more suitable for lake preparation, while desorption occurred, possibly due to crystalline transformation of $CaCO_3$. The isothermal analysis and model fitting results suggested that the β-carotene and $CaCO_3$ particles combined via a monolayer adsorption process driven by physical force. Electrostatic attraction likely participated in the process due to the net negative surface charges of β-carotene dispersion and positively charged groups on the $CaCO_3$ particle surfaces. Ethanol, ultrasonic treatment, and drying method significantly influenced the immobilization efficiency (IE) of β-carotene in the lake and light stability of the lake, without affecting its crystal form. The thermogravimetric analysis (TGA) and differential scanning calorimetry (DSC) curves confirmed absorption of β-carotene onto $CaCO_3$. Fourier-transform infrared spectroscopy (FTIR) and X-ray photoelectron spectroscopy (XPS) analyses indicated no obvious chemical bonds between β-carotene and $CaCO_3$. Energy-dispersive spectroscopy (EDS) confirmed the presence of β-carotene on surfaces but not in the interior of the $CaCO_3$ particles. The adsorption of β-carotene by calcium carbonate was further confirmed to be a physical adsorption on surface.

**Keywords:** β-carotene; colorant lake; calcium carbonate; kinetic adsorption; isothermal adsorption; physical adsorption

## 1. Introduction

The application of natural ingredients in food production is gaining increasing attention recently due to their superior safety, nutritional functionalities, and biological properties compared to synthetic materials [1–3]. As a typical example, β-carotene is favored as a pigment for food use due to its versatile biological functionalities including antitumor, antiaging, antioxidative, free radical scavenging properties, improving immunity, promoting embryo development, and maintaining vision [4]. β-carotene is an oil-soluble pigment, which covers the orange to red chroma depending on its concentration. It is widely used in oily food products, such as butter, margarine, refined fish pulp products, and instant noodles. β-carotene is composed of 11 conjugated double bonds and two ionone rings at both ends. Its structure is extremely unstable, and it is prone to degradation and isomerization in the presence of light, oxygen, free radicals, high temperatures, and metal ions [5]. Encapsulation technologies, including emulsion [6], β-cyclodextrin [7], microcapsules [8], liposomes [9], and solid lipid nanoparticles [10], have been extensively employed to encapsulate β-carotene. These methods aim to coat β-carotene to the maximum extent to achieve the most efficient protection and successively increase its biological activity, which is widely achieved in the literature. Zehra et al. investigated the inclusion of β-carotene by cyclodextrins, which protected β-carotene from UV-mediated oxidation and ensured its biological activity [11]. Alma et al. prepared β-carotene-chitooligosaccharides complexes by mechanochemical methods, and the complexation did not affect the free radical scavenging activity of β-carotene and ensured the water solubility of β-carotene [12]. However, the forementioned encapsulation technologies and their aims are inconsistent with the dyeing purpose that requires adequate β-carotene exposure. Lake pigment technology is more suitable to protect colorants and their dyeing ability. However, to date, only water-soluble pigments have been transformed into colorant lakes in the literature, necessitating the exploration of efficient technology to prepare colorant lakes for oil-soluble pigments, such as β-carotene.

Aluminum hydroxide is the traditional, most common adsorbent for food-grade colorant lake preparation [13]. However, in recent years, several studies have shown that excessive aluminum accumulation in the body causes aluminum poisoning and aggregation in the brain, liver, spleen, kidneys, thyroid, and other tissues and organs. Only 10–15% of consumed aluminum is excreted. The remainder accumulates in the body and interacts with proteins, enzymes, and other components to affect various biochemical reactions [14]. Aluminum deposition in brain and nerve cells causes memory loss, mental decline, and slow reactions, possibly accelerating the aging process and inducing neurodegenerative diseases, such as Alzheimer's disease [15], Parkinson's disease [16], dialysis encephalopathy, and amyotrophic lateral sclerosis [17]. Aluminum deposition in bones can lead to osteoporosis, fractures, and joint pain, while deposition in skin can reduce skin elasticity and increase wrinkles [18–20]. Therefore, the current food additive standard of China, <Standard for the Use of Food Additives> (GB 2760-2014) [21], has reduced the application range and abolished some aluminum additives, compared with the previous version (GB 2760-2011) [22,23]. For example, the application of indigo carmine aluminum hydroxide lake has been restricted. Therefore, our previous study developed a calcium-carbonate-based colorant lake as a substitute for the aluminum-hydroxide-based lake [24]. $CaCO_3$ presents no health risk and offers additional benefits as a calcium supplement. Previous studies have used $CaCO_3$ as an adsorbent due to its multiple surface modifiability and significant surface charge [25–27]. Studies on indigo carmine have shown that coordination bond formation between sulfonic acid groups and calcium ions can form $Ca^{2+}$-IC complexes [28]. Subsequent analysis of $CaCO_3$ and *Monascus* red pigments (MPs) binding indicated physical attraction as the main driving force behind MPs and $CaCO_3$ complex formation, most likely the electrostatic attraction between $Ca^{2+}$ and glutamate residues in MPs. MPs-$CaCO_3$ complex formation is a multistep process involving adsorption, aggregation, crystallization, and desorption [24].

A preliminary experiment in our laboratory confirmed that coprecipitation could be used to form an ideal β-carotene-$CaCO_3$ colorant lake. Although aluminum hydroxide could precipitate with β-carotene, the color intensity and yield of the precipitation were relatively low, making it unsuitable for producing colorant lakes for oil-soluble pigments. The performance variation of the two substrates when constructing lakes for oil-soluble pigments could be due to their microstructural disparity, which will be further explained in future studies.

This study successfully prepared a β-carotene-$CaCO_3$ colorant lake for the first time, which was characterized in terms of light stability, IE, and morphological properties. The influence of ethanol, ultrasonic treatment, and drying method on the colorant lake were investigated. Additionally, the kinetic and isotherm adsorption curves were examined to assess the $CaCO_3$ and β-carotene interaction in the reaction solution, while the existing states of the β-carotene in the lake were analyzed via ultraviolet/visible spectroscopy (UV/Vis), Fourier-transform infrared spectroscopy (FTIR), thermogravimetric analysis

(TGA), differential scanning calorimetry (DSC), X-ray photoelectron spectroscopy (XPS), and energy-dispersive spectroscopy (EDS).

## 2. Materials and Methods

### 2.1. Subsection

Natural β-carotene (96% *w*/*w*) and chemicals of analytical grade, including calcium chloride (96% *w*/*w*), sodium carbonate (99.5% *w*/*w*), sodium hydroxide (99% *w*/*w*), and n-hexane (97% *w*/*w*), were obtained from Macklin Inc. (Shanghai, China). The ethanol (99.7% *w*/*w*) and hydrochloric acid (37% *w*/*w*) were supplied by Beijing Chemical Plant (Beijing, China). A laboratory water purification system (HYP-QX-UP, Huiyipu Ltd., Beijing, China) was employed to produce the deionized (DI) water used for the solutions required for the experiment.

### 2.2. Preparation of the β-Carotene-$CaCO_3$ Colorant Lake

The β-carotene-$CaCO_3$ colorant lakes were prepared via a suspension–precipitation method. Here, 2.0 g of sodium carbonate was dissolved in a 198.0 g water/ethanol solution, while 2.1 g of calcium chloride was dissolved in a 25.0 g water/ethanol solution. The pH of the solutions was adjusted to 9.3 using 1 mol/L HCl and NaOH solutions. The two solutions were heated and kept in a water bath at 40 °C before use. Next, 50.0 mg of β-carotene was added to the sodium carbonate solution, followed by shearing at 18,000 r/min for 90 s (IKA ULTRA-TURRA ® T25 digital, Staufen, Germany) to form a suspension solution with a β-carotene. Then, the calcium chloride solution was mixed with the suspension solution while shearing at 18,000 r/min for 150 s. Since the high-speed shearing inevitably increased the solution temperature, an ice-water bath was employed to control the temperature below 45 °C. After shearing, the reaction solutions were centrifuged at 5000 r/min for 15 min to obtain the $CaCO_3$ precipitates adsorbed with pigments. The sediment was dried in a vacuum drying oven at 45 °C for 12 h and ground in a mortar to acquire an edible β-carotene-$CaCO_3$ colorant lake.

### 2.3. Quantification of the β-Carotene in the Lake

Here, 50.0 mg of the colorant lake was dispersed into a 1 mL acid solution with a hydrochloric acid concentration of 1 mol/L for complete dissolution. Then, 3 mL of an organic solvent (1 mL of ethanol + 2 mL n-hexane) was added and mixed, followed by shaking using a vortex shaker for 45 s to dissolve the β-carotene into the organic phase. The mixture was left to stand to allow complete phase separation, after which the upper layer (solvent phase) was collected. The water phase (lower phase) was extracted by repeating this process twice more, after which the solvent phases of the three extractions were combined and filtered through 0.22 μm organic membranes. The absorbance of the filtered solution was measured at 450 nm using a spectrophotometer (Shimadzu UV3600Plus, Kyoto, Japan). High-purity β-carotene (99% *w*/*w*) was used to establish the standard linear curve of the β-carotene at 450 nm absorbance. Then, the amount of β-carotene immobilized by calcium carbonate was calculated and denoted as $M_1$ (mg), and the immobilization efficiency (IE) of β-carotene (mg/g) was calculated using Equation (1):

$$IE = M_1/M_2 \times 100\% \quad (1)$$

where $M_1$ is the mass of the β-carotene detected in the lake (mg), and $M_2$ is the mass of the lake, which is 0.05 g.

### 2.4. Kinetic Adsorption Analysis

(1) The influence of temperature: The suspension and calcium chloride solutions were prepared as described in Section 2.2. The solutions were preheated in a water bath at 30 °C before mixing and shearing at 18,000 r/min for 90 s. Next, a magnetic stirrer with a temperature-control function was used to stir the reaction solutions at 30 °C. At predetermined time points, the reaction solution was centrifuged, as described in Section 2.2,

to obtain the sediment. The experiment lasted 48 h (2880 min). The sediment collected at each time point was dried in a vacuum drying oven at 45 °C for 12 h and ground in a mortar to obtain the $CaCO_3$ lake. The quantity of β-carotene in the lake ($M_1$) was measured using the method described in Section 2.3, while the adsorbed β-carotene at each time point ($q_t$) was calculated using Equation (2). The $q_t$ was plotted against time to obtain the adsorption kinetic curves, which were also determined at 40 °C and 50 °C using the same method employed for the 30 °C measurements but with modifications regarding the water bath temperature and stirring parameters.

$$q_t = M_1/M_c \times 100\% \tag{2}$$

where $M_1$ is the mass of the β-carotene detected in the lake (mg), and $M_c$ is the mass of the $CaCO_3$ in the lake (g).

(2) The influence of pH: The kinetic curves were determined using the same method described above, with modifications regarding the temperature, which was set at a constant 40 °C, while pH values of 8, 9.3, and 10 were used to investigate the influence of pH on the kinetic curve of lake formation.

*2.5. Isothermal Adsorption Analysis*

To prepare the working solutions, 0.80 g of $Na_2CO_3$ was dissolved in 79.2 g of DI water. NaOH/HCl was used to adjust the solution pH to 9.3, after which 0.84 g of $CaCl_2$ was dissolved in 10.0 g of DI water to prepare the $CaCl_2$ for later use. The method described in Section 2.2 was used to prepare a series of lakes by adding different concentrations of β-carotene, i.e., 0.5 mg/mL, 1 mg/mL, 1.5 mg/mL, 2 mg/mL, 3 mg/mL, 4 mg/mL, 5 mg/mL, 6 mg/mL, 7 mg/mL, 8 mg/mL, 9 mg/mL, and 10 mg/mL. The β-carotene content in the lakes was measured using the method described in Section 2.3, while adsorbed β-carotene quantity per gram of $CaCO_3$ at equilibrium ($q_e$ (mg/g)) was calculated using Equation (2). The $q_e$ is actually the qt at equilibrium stage. The unabsorbed β-carotene concentration in the reaction solution ($C_e$ (mg/mL)) was calculated using Equation (3). The isothermal adsorption curve was drawn by plotting $q_e$ against $C_e$:

$$C_e = \frac{C_1V - q_e \times m_{CaCO_3}}{V} \tag{3}$$

where $C_1$ is the concentration of β-carotene added to the solution, $C_e$ is the concentration of unabsorbed β-carotene at equilibrium (mg/mL), $m_{CaCO_3}$ is the amount of calcium carbonate produced (g), and V is the solution volume (mL).

To clarify the adsorption process between the $CaCO_3$ and β-carotene, the isothermal adsorption data were fitted into Langmuir and Freundlich isothermal adsorption curve equations, as shown in Equations (4)–(6):

$$\text{Langmuir}: \frac{C_e}{q_e} = \frac{1}{q_{max} \times K_L} + \frac{C_e}{q_{max}} \tag{4}$$

$$R_L = \frac{1}{1 + (K_L \times C_0)} \tag{5}$$

$$\text{Freundlich}: \ln(q_e) = \ln(K_f) + \frac{\ln(C_e)}{n} \tag{6}$$

where $q_{max}$ is the maximum value of $q_e$ (mg/g), $K_L$ is the adsorption equilibrium constant of the Langmuir model (L/mg), n is the Freundlich constant, representing adsorption strength, $K_f$ is the adsorption equilibrium constant of the Freundlich model ($mg^{1-1/n} \cdot g^{-1} \cdot L^{1/n}$), and $C_0$ is the initial β-carotene concentration in solution after each addition (mg/L).

### 2.6. Preparation of Lake with Varying Ethanol Contents, Ultrasonic Treatments, and Drying Methods

(1) Varying ethanol content: The lake was prepared using the method described in Section 2.2. The ethanol volume ratio in the water/ethanol solution were set to 0% (water: ethanol = 1:0), 10% (water: ethanol = 9:1), 20% (water: ethanol = 4:1), 30% (water: ethanol = 7:3), 40% (water: ethanol = 3:2), and 50% (water: ethanol = 1:1), respectively.

(2) Varying ultrasonic treatments: The lake was prepared using the method as described in Section 2.2, which was modified by adding ultrasonic treatment (Kunshan Shumei KQ2200DE, Kunshan, China) after the second shearing operation at 2 min, 4 min, 6 min, 8 min, and 10 min, respectively.

(3) Varying drying methods: The lake was prepared using the method described in Section 2.2, which was modified by employing freeze drying, vacuum drying, and oven drying, respectively, to dry the sediment.

### 2.7. Determination of the Light Stability

The lake sample was placed in a light incubator (Shanghai shbo-xun SPX-150B-Z, Shanghai, China) with stable, simulated natural light illumination at 12,000 LX and a temperature of 25 °C. The CIE L*a*b* color scale was used for evaluation. A color difference meter (Konica Minolta CM-3610A, Tokyo, Japan) was used to determine the change in redness ($\Delta a^*$) and total color difference ($\Delta E$) after 48 h. Lower $\Delta a^*$ and $\Delta E$ absolute values indicated higher color stability.

### 2.8. Microscopic Morphology Analysis

The microscopic morphology of the β-carotene-$CaCO_3$ colorant lake was observed using SEM. The lake was dried in an oven at 45 °C for 12 h to minimize its moisture content before observation. The dried powder samples were sputtered with gold, and their microscopic morphology was observed using a microscope (Hitachi S-4800, Tokyo, Japan). The prepared lake was placed in a mortar, and liquid nitrogen was poured into the mortar to freeze the lake sample, which was then quickly ground. The C, Ca, and O distribution on the surfaces and cross-sections of the complex lake particles was analyzed via EDS using point-scanning mode.

### 2.9. Crystal Structure Analysis

The crystal morphology of the $CaCO_3$ lake was analyzed via X-ray diffraction. The sample was scanned at a speed of 2°/min in a range of 5 to 90° (2θ) using the fine scanning mode of a diffractometer (Rigaku SmartLab SE, Tokyo, Japan). The scan data were analyzed using MDI Jade 6.

### 2.10. Zeta Potential Measurement

The lake was dispersed in DI water, after which about 1 mL of the solution was injected into a capillary zeta potential cell using a syringe. The cell was placed in a Malvern Zetasizer Nano ZS90 potentiometer (Malvern Instruments, Malvern, UK) to determine its surface charge, while the potential was measured at ambient temperature. The zeta potential of the β-carotene particles was immediately determined after the shearing operation to disperse β-carotene into the sodium carbonate solution, as shown in Section 2.2.

### 2.11. Particle Size Measurement

A Malvern Mastersizer 2000 laser particle size analyzer (Malvern Instruments, Malvern, UK) was used to determine the particle sizes at 25 °C, and the volume-weighted average diameter, $d_{4,3}$, was recorded.

### 2.12. BET Surface Area Measurement

This process was performed using a BET machine (Quanta Autosorb-iQ, Houston, TX, USA). Before the measurement, the lake samples were ground to obtain smaller particles.

The lake and $CaCO_3$ powders were degassed at 80 °C, and the $N_2$ adsorption data were determined and analyzed.

### 2.13. TGA and DSC Analyses

A synchronous thermal analyzer (TA Instruments Q600, New Castle, DE, USA) was used for the TGA and DSC analyses. The lake samples were placed on a standard aluminum pan and sealed tightly with a perforated aluminum sheet. The samples were heated from room temperature to 800 °C at a constant rate of 10 °C/min using nitrogen as a gas atmosphere. The weight change and heat flow signal were recorded and analyzed [29].

### 2.14. FTIR Analysis

The β-carotene-$CaCO_3$ colorant lake was dried in a vacuum drying oven at 45 °C for 12 h before measurement. The sample was mixed with potassium bromide at a mass ratio of 1:150 and pressed into pellets for analysis. The light transmittance was determined in a range of 400 $cm^{-1}$ to 4000 $cm^{-1}$ at a resolution of 2 $cm^{-1}$ using a Nicolet iS5 FTIR spectrometer (Thermo Fisher, Waltham, MA, USA).

### 2.15. XPS Analysis

The prepared colorant lake was dried in a vacuum drying oven at 45 °C for 12 h before measurement. XPS analysis was performed using an Escalab 250Xi+ X-ray photoelectron spectrometer (Thermo Fisher, Waltham, MA, USA), and C, O, and Ca levels were analyzed.

### 2.16. Statistical Analysis

The data were processed and graphed using Origin 2021. All assays were performed in triplicate, and the results are presented as mean ± standard deviation (SD). Significance analysis was performed using IBM SPSS Statistics 26, while differences were regarded as significant when $p < 0.05$.

## 3. Results and Discussion

### 3.1. Kinetic Adsorption Curve

Since β-carotene is suspended as small particles in aqueous solutions rather than being fully dissolved, the process of combining it with $CaCO_3$ does not involve typical adsorption. Assuming that a β-carotene suspension particle was a whole entity involved in the adsorption process, standard adsorption evaluation methods could be employed, such as kinetic and isothermal curve analyses.

Figure 1a,b illustrate the kinetic curves. The temperature variation from 30 °C to 50 °C distinctly influenced the kinetic curves (Figure 1a). Minimal $q_t$ variation was evident at 30 °C, from about 7.827 mg/g to 10.928 mg/g in the first 30 min, indicating stable β-carotene and $CaCO_3$ binding. From 30 min to 2880 min, the $q_t$ fluctuation range increased from 6.696 mg/g to 11.899 mg/g. At 40 °C, the $q_t$ displayed an initial increase to a maximum value of 16.731 mg/g at 5 min, followed by a slight decrease over 30 min, after which the $q_t$ fluctuated in a range of 6.545 mg/g to 13.388 mg/g. At 50 °C, the $q_t$ remained almost constant during the first 6 min at about 15 mg/g, then gradually decreased to around 10 mg/g at 30 min and approximately 2 mg/g at the end of the experiment. The $q_t$ values of all the curve peaked during the first 10 min of the adsorption reaction, after which they generally declined until the end of the experiment. This was especially evident for the curve at 50 °C. This was consistent with reports on an MPs-$CaCO_3$ lake [24]. The $q_t$ decrease indicated β-carotene desorption from the $CaCO_3$ particles, which was related to the crystalline $CaCO_3$ transformation from an amorphous form to calcite. In summary, a medium temperature (40 °C) increased the β-carotene content in the lake compared to a lower temperature (30 °C) and facilitated higher adsorption stability (less desorption) compared to a higher temperature (50 °C). Therefore, 40 °C was selected for lake preparation in the subsequent experiment.

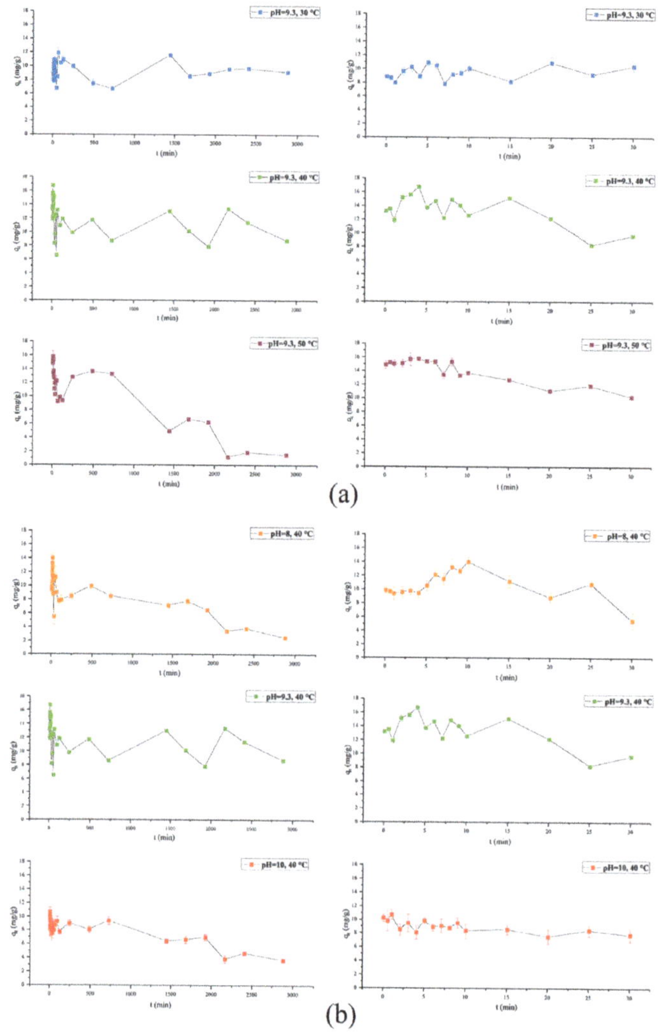


**Figure 1.** The influence of temperature (**a**) and pH (**b**) on the kinetic adsorption curve (left: 0–2880 min, right: 0–30 min).

Figure 1b shows the influence of pH on the kinetic curves. The pH can affect the functional groups, charge level, and ion form of the adsorbent, consequently impacting its adsorption capacity [30]. At pH 8 and 9.3, the $q_t$ exhibited an initial increase, reaching the maximum levels, followed by a decline. However, at pH 10, the maximum level appeared at the beginning of the measurement, after which the $q_t$ decreased throughout the remainder of the test. The maximum qt values at pH 8, 9.3, and 10 were 14.008 mg/g, 16.731 mg/g, and 10.745 mg/g, respectively, indicating that pH 9.3 was more suitable for lake preparation. A previous study involving the preparation of an MPs-$CaCO_3$ lake also investigated the pH of the reaction solution, confirming that pH 10.5 was optimal due to the highest $q_t$ level [24]. The pH values of achieving maximum of $q_t$ for MPs and β-carotene are different but close. The difference reflects the disparity between the two pigments in terms of chemical composition, and the proximity of the pH values may suggest similar driving forces or manners existing in the adsorption processes of the two pigments onto $CaCO_3$. Since the

*Monascus* pigments were negatively charged due to their carboxyl groups in a basic pH range, they could be absorbed by $CaCO_3$. The $CaCO_3$ surface charge is determined by the presence of -$Ca^+$ and -$CO_3^-$ sites on its surface and their possible hydrolysates, such as -CaOH, -Ca $(OH_2)^+$, -$CO_3H$, and -$CO_3$ $(OH_2)^-$ [31]. The zeta potentials of β-carotene suspension, calcium carbonate particles, and lake particles (β-carotene-$CaCO_3$ complex particles) with different pH values to those in the preparation process are shown in Table 1. It is evident that all of the three particles exhibit negative charges on surfaces. The zeta potential of calcium carbonate was higher than that of the color lake and β-carotene suspension. The results indicated that the presence of -$Ca^+$ site on the surface of calcium carbonate and its possible cationic hydrolysate adsorbed pigment suspensions that were negatively charged, resulting in more net negative charges on the surface of the complex lake particles [24,32]. The results verified the important role of electrostatic attraction as a driving force for the formation of β-carotene-$CaCO_3$ complex particles.

**Table 1.** The zeta potential of different particles in reaction solution.

| | pH = 8 | pH = 9.3 | pH = 10 |
|---|---|---|---|
| β-carotene-$CaCO_3$ complex particle | $-26.1 \pm 0.51^{b}$ | $-27.7 \pm 0.47^{b}$ | $-27.3 \pm 0.81^{b}$ |
| $CaCO_3$ particle | $-17.1 \pm 0.69^{a}$ | $-18.1 \pm 0.54^{a}$ | $-17.6 \pm 0.41^{a}$ |
| β-carotene suspension | $-25.8 \pm 0.20^{b}$ | $-28.3 \pm 0.26^{b}$ | $-27.5 \pm 0.35^{b}$ |

Values not sharing the same superscript letters in one column are significantly different ($p < 0.05$).

### 3.2. Isothermal Curve of Adsorption

Both of the $q_e$ and $C_e$ values increased after adding β-carotene to the reaction solution (Figure 2a). The $q_e$ increased rapidly and exhibited a steep slope as the $C_e$ rose from 0 mg/mL to 6 mg/mL. At levels higher than 6 mg/mL, the slope flattened abruptly and leveled off. When the added β-carotene quantity exceeded 7 mg/mL, the $CaCO_3$ particles were saturated, and the $q_e$ reached a maximum value of about 100.02 mg/g, indicating that every gram of $CaCO_3$ could immobilize 100.20 mg of β-carotene.

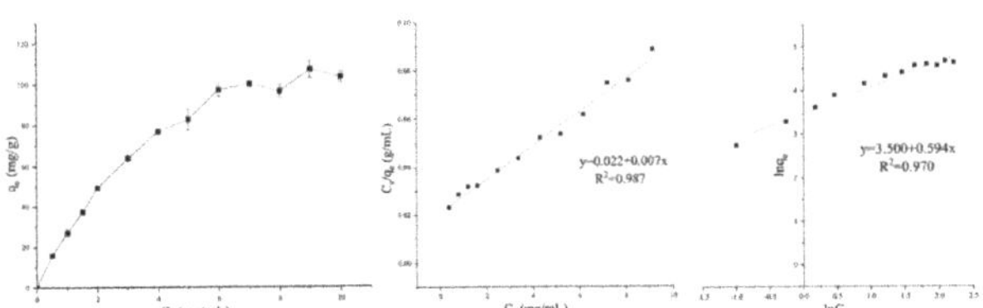

(a) (b) (c)

**Figure 2.** The isothermal adsorption curves of the β-carotene-$CaCO_3$ colorant lake (40 °C) (**a**), linear Langmuir equation data fitting (**b**), and linear Freundlich equation data fitting (**c**).

Figure 2b,c show the fitting of isothermal data into the linear Langmuir and Freundlich equations. The $R^2$ values indicated that the data fitted both of the Langmuir and Freundlich models well. However, the Langmuir model displayed slightly better correlations. The $q_{max}$, $K_L$, and $R_L$ values of the Langmuir model were calculated as 142.86 mg/g, 0.32 mL/mg, and 0.24~0.86, respectively, using Equations (4) and (5). According to the Langmuir adsorption theory, the adsorption is a monolayer adsorption process, and when the $R_L$ value is less than 1, the adsorption process is favorable [33]. For the Freundlich model, $K_f$ was calculated to be 33.12, and n was calculated to be 1.68. According to the Freundlich adsorption theory, when $n > 1$, adsorption is generally a physical process [24,34]. The results suggest that the adsorption between the β-carotene droplets and $CaCO_3$ is likely a monolayered process driven by physical molecular forces. The conclusion was partially consistent with the isothermal MPs-$CaCO_3$ absorption analysis, which agreed with the Freundlich model but did not fit into the Langmuir model [24]. Compared with the dissolved *Monascus* pigment molecules, the suspended β-carotene particles were not expected to form multiple adsorption layers on the $CaCO_3$ particle surfaces due to their significant size.

### 3.3. Influence of Ethanol on the β-Carotene-$CaCO_3$ Lake Preparation

Studies have shown that the appearance of ethanol significantly influences $CaCO_3$ formation and crystalline transformation [35,36]. Therefore, it is necessary to clarify the effect of ethanol on the properties of $CaCO_3$-based colorant lake. Figure 3a shows the impact of the ethanol concentration in the lake-forming solution on the β-carotene IE. Due to the poor solubility of $Na_2CO_3$ and $CaCl_2$ in solutions containing over 50% ethanol, only lower ethanol concentrations were examined. Ethanol addition significantly enhanced the β-carotene IE, reaching maximum levels after adding 20% ethanol, while the IE fluctuated at higher ethanol concentrations. Small and uniform size could facilitate the adsorption of particles [37], while the introduction of low-concentration ethanol resulted in the reduction in pigment particle size and the enhancement of absorbability (Figure 3a). Figure 3b shows the effect of the ethanol concentration on the micromorphology of the lake. The $CaCO_3$ in the lake displayed a clear outline with flat and smooth surface, exhibiting typical calcite topography as reported in the literature [38]. The XRD analysis (Figure 3c) confirmed that only one type of crystal form of $CaCO_3$, calcite, was present in the lake, regardless of the ethanol concentration. In addition to calcite, the SEM image showed some particles that appeared irregular in shape but displayed a wide range of sizes, which were likely immobilized β-carotene particles. The β-carotene dispersed on the $CaCO_3$ particle surfaces in both the ethanol-added and ethanol-free lakes. As the ethanol content increased, a decrease in particle size and enhanced dispersion of particles were observed (Figure 3a). The findings showed that, even with the coprecipitation approach, there was a chance that the $CaCO_3$ particles and β-carotene formed sediment due to various attraction forces rather than physical entanglement or direct physical entrapment, since a large amount of β-carotene appeared on the surface of $CaCO_3$ particles.

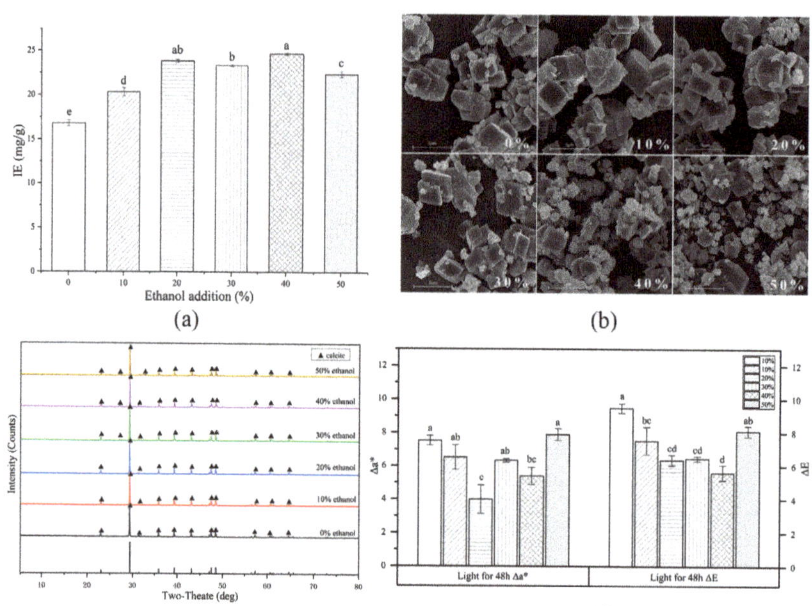


**Figure 3.** The influence of ethanol on the IE (**a**), micromorphology (**b**), crystalline form (**c**), and light stability (**d**) of the β-carotene-$CaCO_3$ lake. (Values not sharing the same superscript letters in (**a**) and (**d**) are significantly different ($p < 0.05$).)

In light conditions, the Δa* value decreased as the ethanol concentration increased from 0% to 20%; however, beyond this threshold, the Δa* exhibited an increasing trend, with the minimum value evident at an ethanol concentration of 20%. Similarly, the ΔE initially decreased, followed by an increase at a higher added ethanol concentration. These findings demonstrated that lower ethanol concentrations enhanced the stability of the β-carotene-$CaCO_3$ colorant lake, while higher concentrations detrimentally affected its stability.

### *3.4. Influence of Ultrasonic Treatment on the β-Carotene-$CaCO_3$ Lake Preparation*

It is worth investigating the effect of ultrasonic treatment on the β-carotene-$CaCO_3$ lake properties since it can potentially improve $CaCO_3$ particle uniformity [39]. Figure 4a shows the effect of ultrasonic treatment on the β-carotene IE. The lakes produced with varying ultrasonic treatment durations did not exhibit statistically significant IE variation (approximately 16.5 mg/g). This was significantly lower than the lakes prepared without ultrasonic treatment but with medium ethanol concentrations (Figure 3a) ($p < 0.05$). The $CaCO_3$ particles in the lakes prepared via different ultrasonic treatments showed no distinct differences in size and morphology, and had similar size and morphology to those illustrated in Figure 3a, but exhibiting significantly fewer irregular particles, which was consistent with the lower β-carotene IE in the ultrasonically treated lake.

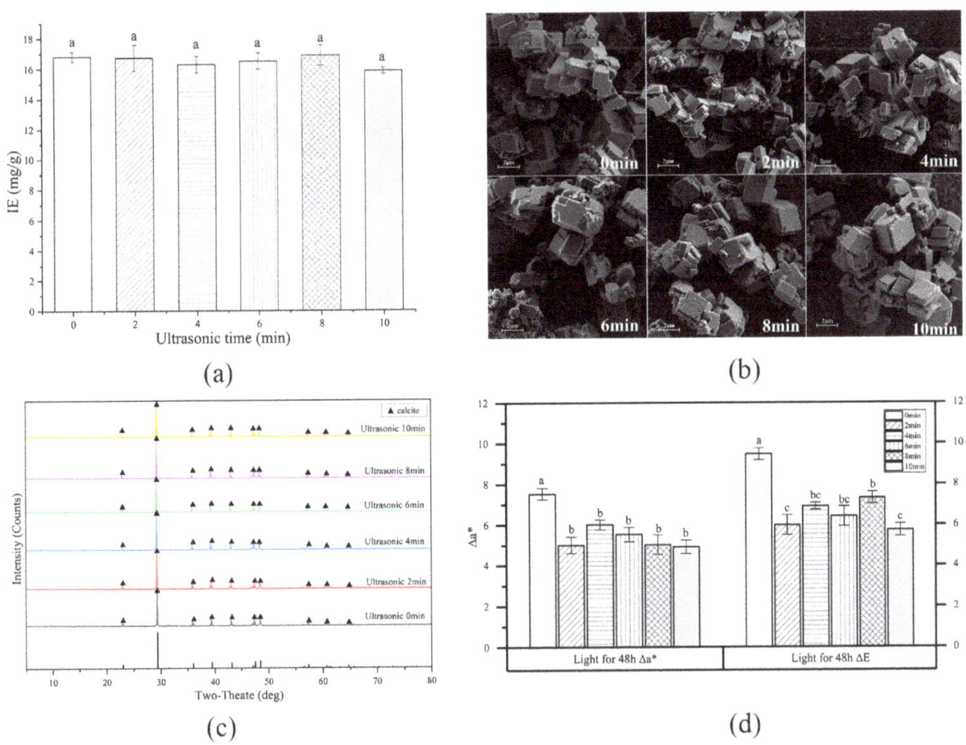


**Figure 4.** The influence of ultrasonic treatment on the IE (**a**), micromorphology (**b**), crystalline form (**c**), and light stability (**d**) of the β-carotene-$CaCO_3$ lake. (Values not sharing the same superscript letters in (**a**) and (**d**) are significantly different ($p < 0.05$).)

Figure 4d shows light stability of the β-carotene-$CaCO_3$ lake prepared with different ultrasonic durations. The Δa* and ΔE values in the ultrasonic group after 48 h illumination were lower than those without ultrasonic treatment, showing that ultrasound significantly improved the colorant lake stability. Under illumination, longer ultrasonic treatment did not further improve the light stability. The results confirmed that ethanol addition was more efficient than ultrasonic treatment in improving the light stability since the former yielded the lowest Δa* value (about 4) among all the tests.

### *3.5. Influence of Drying Methods on β-Carotene-$CaCO_3$ Lake Preparation*

Figure 5a shows the IE of the β-carotene colorant lake prepared using different drying methods. The IE of the colorant lake prepared using vacuum drying was the highest (16.809 mg/g), followed by the sample prepared via freeze drying (15.819 mg/g), while that of the colorant lake prepared using oven drying was the lowest (8.562 mg/g). The results indicated that oven drying destroyed a large proportion of the immobilized β-carotene in the β-carotene-$CaCO_3$ lake due to oxidation in the high-temperature, oxygen-rich environment. The IE discrepancy between vacuum and freeze drying was minimal (<1 mg/g) and not significant ($p > 0.05$). Figure 5b presents the micromorphology of the β-carotene-$CaCO_3$ colorant lakes prepared in different drying conditions. The particles in the three lakes were similar, showing aggregation, size uniformity, and typical calcite topography. The XRD analysis confirmed that only calcite was present in the three kinds of lakes (Figure 5c). However, freeze drying demands higher energy input, more complicated operation, and a longer waiting time compared to vacuum drying. Therefore, vacuum drying was more suitable for colorant lake preparation.

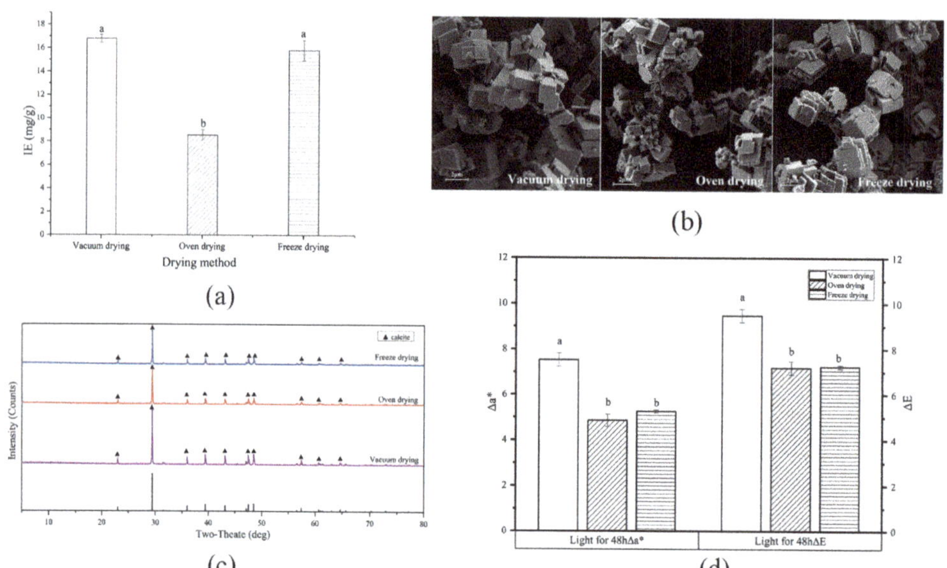


**Figure 5.** The influence of drying methods on the immobilization efficiency (IE) (**a**), micromorphology (**b**), crystalline form (**c**), and light stability (**d**) of the β-carotene-$CaCO_3$ lake. (Values not sharing the same superscript letters in (**a**) and (**d**) are significantly different ($p < 0.05$).)

Figure 5d shows the light stability of the lake products exposed to different drying methods. In light conditions, the Δa* and ΔE values of the sample prepared using vacuum drying were higher than samples prepared using oven drying and freeze drying, which had no significant difference between them ($p > 0.0.5$). However, the β-carotene IE of the colorant lake prepared with oven drying (8.562 mg/g) was about half that prepared with freeze drying (15.819 mg/g) and vacuum drying (16.809 mg/g). It stands to reason that, for the lake prepared by oven drying, its much lower IE contributed to its superiority in light stability. Thus, for all the drying methods, freeze drying is the best, considering its excellent performances in both IE and light stability. However, compared with freeze drying, vacuum drying has the characteristics of small equipment investment, large processing capacity, and simple operation [40]. For economical consideration, the vacuum drying, which only showed slightly insufficient light stability of prepared lake compared to the freeze drying method in the test, is also a good choice.

In summary, ethanol addition, ultrasonic treatment, and drying methods all had no significant influence on the micromorphology and crystalline form of the lake. Ethanol addition at suitable concentration is an efficient way to improve IE of β-carotene in the lake, ultrasonic treatment had no obvious influence on the IE, and oven drying could distinctly decrease the IE in the lake. Compared to the freeze drying, the vacuum drying method supplied a compromised choice with a little impaired light stability but expected economic benefit. On the one hand, foods are stored in light-proof packaging until consumed in some cases; thus, high light stability of the foods' color is not a key requirement, for example, the case of using the β-carotene-$CaCO_3$ lake to dye dietary supplement tablets or biscuits. On the other hand, foods need to be exposed to light for display purposes in some cases; thus, high light stability is required, for example, the case of using the β-carotene-$CaCO_3$ lake to dye breads or cakes. Thus, in different scenarios, production parameters of the β-carotene-$CaCO_3$ lake could be adjusted to make an economical choice between color intensity (IE) and light stability.

### 3.6. Zeta Potential, Particle Size, and Surface Area

The characteristics of dried calcium carbonate, lake, β-carotene, and β-carotene suspension were compared, and the results were shown in Table 2. There were significant differences in zeta potential and diameter of the four kinds of particles ($p < 0.05$). The diameter of lake particles (8.05 μm) was significantly higher than that of calcium carbonate particles (6.85 μm) due to the adsorption of β-carotene, which had an initial diameter of 102.82 μm and a significantly reduced diameter of 0.89 μm after shearing. The results confirmed that shearing operation largely decreased the size of β-carotene suspension, which is critical for the adsorption process and formation of homogeneous lake product. In the BET results, the surface area of calcium carbonate (0.99 $m^2/g$) was larger than the surface area of the lake (0.90 $m^2/g$) and β-carotene (0.72 $m^2/g$), which was consistent with the particle size results. The BET of β-carotene suspension was large but with much bigger deviation compared to the others, indicating that the suspended particles after shearing had a heterogeneous structure. In the process of MPs adsorption by $CaCO_3$, the particle size of the lake was smaller than that of calcium carbonate and the BET of the lake was larger than that of calcium carbonate [24]. The β-carotene-$CaCO_3$ colorant lake had the opposite result. The different influence of β-carotene and MPs on the size of $CaCO_3$ particles could be ascribed to the distinct existing states of them in reaction solution. The results of zeta potential were consistent with those in Table 1. Calcium carbonate had the highest potential (−21.07 mV), and β-carotene was negatively charged (−28.83 mV). The results suggest that during the formation of the β-carotene-$CaCO_3$ complex particles, the positively charged ions (-$Ca^+$, -Ca $(OH_2)^+$) on the calcium carbonate particle surface adsorbed the negatively charged ions on the surface of β-carotene suspension, resulting in a decrease in the potential of the lake.

**Table 2.** Basic characteristics of calcium carbonate, lake, β-carotene, and β-carotene suspension.

| | **Zeta Potential (mV)** | **Diameter (μm)** | **BET ($m^2/g$)** |
|---|---|---|---|
| $CaCO_3$ | $-21.07 \pm 0.44^{d}$ | $6.85 \pm 0.13^{c}$ | $0.99 \pm 0.08^{a}$ |
| β-carotene-$CaCO_3$ colorant lake | $-24.37 \pm 0.44^{c}$ | $8.05 \pm 0.02^{b}$ | $0.90 \pm 0.03^{a}$ |
| β-carotene | $-28.83 \pm 0.28^{a}$ | $102.82 \pm 5.77^{a}$ | $0.72 \pm 0.04^{a}$ |
| β-carotene suspension | $-27.30 \pm 0.62^{b}$ | $0.89 \pm 0.01^{d}$ | $2.68 \pm 1.29^{b}$ |

Values not sharing the same superscript letters in one column are significantly different ($p < 0.05$).

### 3.7. XPS and EDS Analysis

EDS is an efficient tool to evaluate elemental composition of particles [41]. The SEM images and scanning locations (denoted by black points) of the surface of lake particles (Figure 6a), internal section of lake particles (Figure 6b), surface of $CaCO_3$ particles (Figure 6c), internal section of $CaCO_3$ particles (Figure 6d), β-carotene particles (Figure 6e), and β-carotene suspension particles (Figure 6f) are shown in Figure 6. The atomic percentage and weight percentage obtained through EDS analysis are presented in Table 3. β-carotene (location 6) and β-carotene suspension (location 7) exhibited minimal disparity in the atomic ratios of C and O, which was expected and confirmed that shearing operation did not cause distinct change to the chemical composition of β-carotene. Location 2 was chosen for the irregular particles, which were believed to be β-carotene particles. The findings indicate that the distribution of pigment on the lake's surface is nonuniform. In the calcium carbonate and lake samples, with the exception of location 2, all other locations exhibited atomic percentage of carbon (C) ranging from 17% to 19%, while location 2 displayed an unusually high atomic percentage of C, at approximately 33%. The results indicated that there was similar C element distribution on the surface and internal section of $CaCO_3$ particles in the lake to that of pure $CaCO_3$ particles, which confirmed that there was no β-carotene evenly covering the surface or on the inside of $CaCO_3$ particles. The weight percentage of the samples consistently yielded results, indicating a significantly higher weight percentage of C in location 2. This finding further strengthens the conclusion drawn from the analysis based on weight percentage.

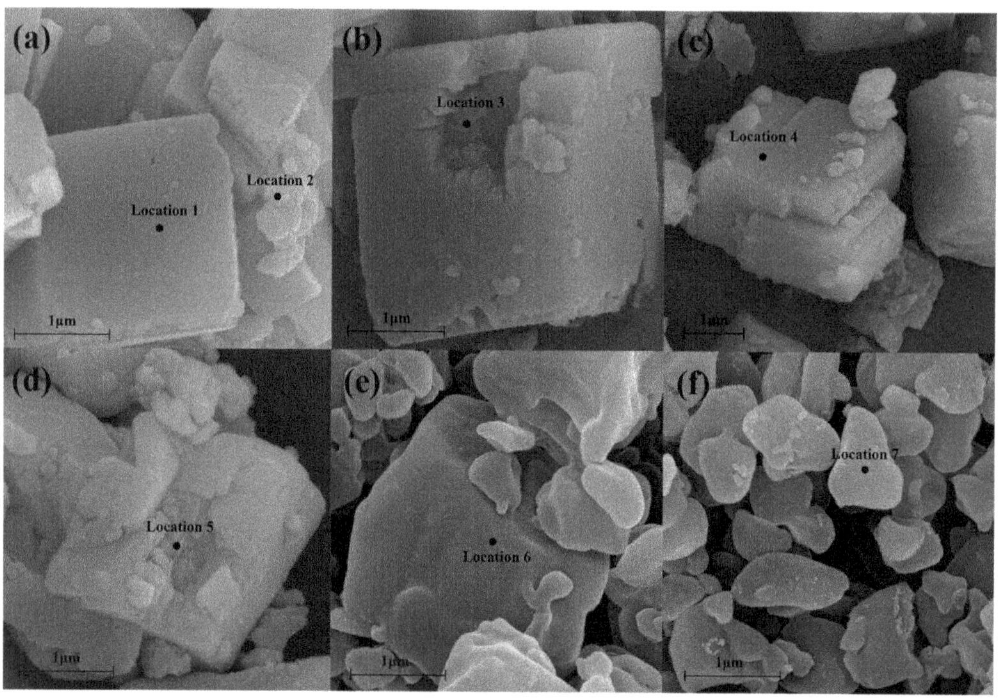


**Figure 6.** The SEM images of the β-carotene-$CaCO_3$ lake (**a**,**b**), the $CaCO_3$ (**c**,**d**), the β-carotene (**e**), and the β-carotene suspension (**f**), and the scanning locations (denoted by black points) for EDS analysis.

**Table 3.** The EDS analysis of the β-carotene-$CaCO_3$ lake and $CaCO_3$.

| | | C | | O | | Ca | | Na | |
|---|---|---|---|---|---|---|---|---|---|
| | | Weight% | Atomic% | Weight% | Atomic% | Weight% | Atomic% | Weight% | Atomic% |
| lake | Location 1 | 11.25 | 17.37 | 59.64 | 69.16 | 29.11 | 13.47 | - | - |
| | Location 2 | 22.82 | 32.73 | 52.73 | 56.76 | 24.45 | 10.51 | - | - |
| | Location 3 | 10.74 | 18.58 | 45.01 | 58.47 | 44.25 | 22.95 | - | - |
| $CaCO_3$ | Location 4 | 11.53 | 17.32 | 63.26 | 71.33 | 25.21 | 11.35 | - | - |
| | Location 5 | 11.55 | 18.80 | 51.79 | 63.31 | 36.66 | 17.89 | - | - |
| β-carotene | Location 6 | 72.84 | 78.13 | 27.16 | 21.87 | - | - | - | - |
| β-carotene suspension | Location 7 | 69.81 | 75.99 | 27.51 | 22.48 | - | - | 2.68 | 1.52 |

XPS uses chemical analysis to characterize surface composition [42]. Figure 7 shows the XPS spectrum analysis results of the particle surfaces. $CaCO_3$ consists of Ca, C, and O, while β-carotene solely comprises C and H. No additional measurable elements were introduced into the β-carotene-$CaCO_3$ complex. Curve fitting indicated that C accounted for 47.98% in the lake and 36.75% of the $CaCO_3$, indicating successful β-carotene adsorption onto the $CaCO_3$ surface. The peaks corresponding to Ca, C, and O in the $CaCO_3$ and color lake were analyzed for comparison. A slight shift was evident among the C, O, and Ca peaks between $CaCO_3$ and the lake. This suggested a physical interaction between the β-carotene and $CaCO_3$, providing further evidence for adsorption of carotene.

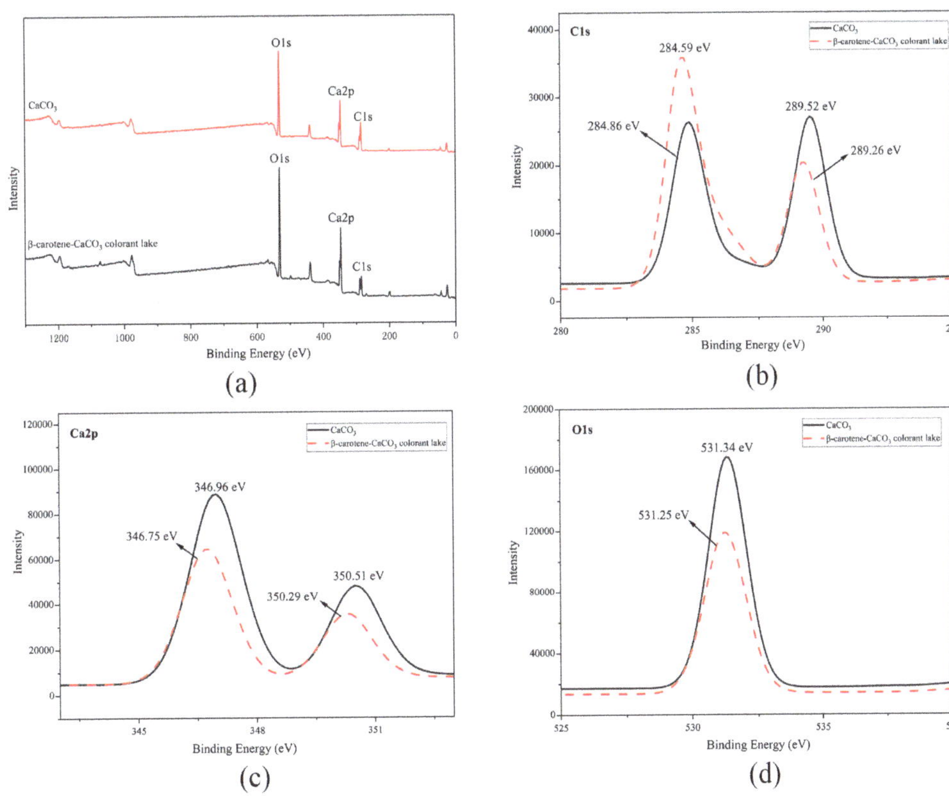


**Figure 7.** The XPS analysis (**a**) and the spectra of different elements (C1s (**b**), Ca2p (**c**), and O1s (**d**)) of the β-carotene-$CaCO_3$ lake and $CaCO_3$.

Combined with the results of XPS and EDS, pigment was attached to the surface of calcium carbonate particles in the lake prepared by coprecipitation method. Calcium carbonate has little or no internal absorption of pigments.

### 3.8. TGA and DSC Analysis

The TGA spectra of pigment, calcium carbonate, and the lake are presented in Figure 8a. The decomposition of the lake occurred in two distinct stages, namely, weight losses near 450 °C and 650 °C, while the decomposition of calcium carbonate only took place around 650 °C. The disparity in weight loss between the lake and calcium carbonate before 650 °C can be attributed to the influence of the pigment on calcium carbonate. The final weight loss of the lake and calcium carbonate was 44% and 41%, respectively, and the weight loss of the lake was slightly greater than that of calcium carbonate [43,44]. When the temperature was near 350 °C, the degradation of the lake was more obvious than that of $CaCO_3$, indicating that β-carotene was attached to $CaCO_3$ particles [45].

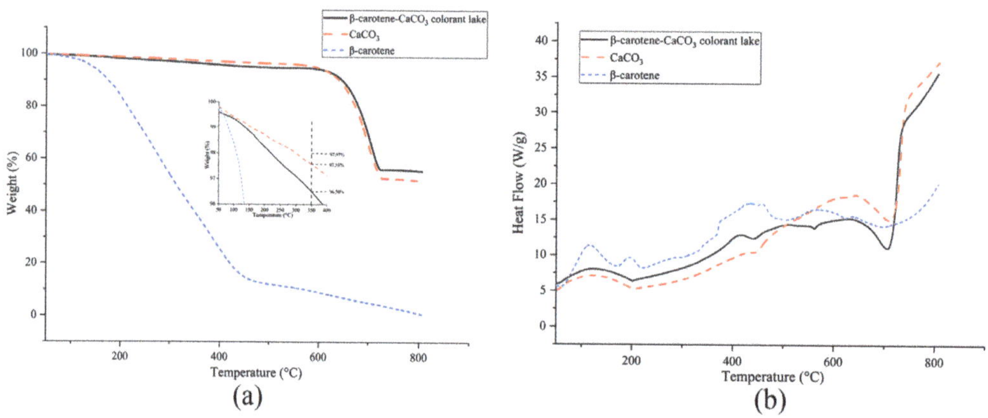


**Figure 8.** TGA (**a**) and DSC (**b**) curves of the β-carotene-$CaCO_3$ lake, $CaCO_3$, and β-carotene.

The DSC curve of the β-carotene (Figure 8b) exhibited significant changes, which was related to its own instability of highly unsaturated structures, while the TGA curve (Figure 7a) showed a considerable decrease in the β-carotene mass as the temperature rose from 50 °C to 425 °C. Previous studies have reported that thermal degradation of β-carotene occurs within the range of 50–150 °C [46], and the thermal cracking takes place at temperatures exceeding 300 °C [47], Consequently, multiple exothermic peaks were observed in the DSC curve. The DSC curve showed that exothermic peaks appeared in both lake and pigment at around 425 °C, while there was no $CaCO_3$ peak at this temperature, which may be due to the thermal cracking of β-carotene. The adsorption of pigment by calcium carbonate was further explained.

### 3.9. FTIR Analysis

Figure 9 illustrates the positions and number of FTIR spectral peaks of the β-carotene, $CaCO_3$, and β-carotene-$CaCO_3$ colorant lakes. Table 4 shows the FTIR spectra peak assignments for $CaCO_3$ and β-carotene. As shown in the figure, except for the peak at 3424.42 $cm^{-1}$, the spectra of $CaCO_3$ and β-carotene-$CaCO_3$ colorant lakes have no distinct difference. The presence of peak at 3424.42 $cm^{-1}$ in the spectrum of the lake is due to the characteristic peak in the β-carotene spectrum at similar wavenumber (3432.39 $cm^{-1}$). No new peak is evident in the spectrum of the lake, confirming that there is no chemical bonding established between β-carotene and $CaCO_3$ in the lake.

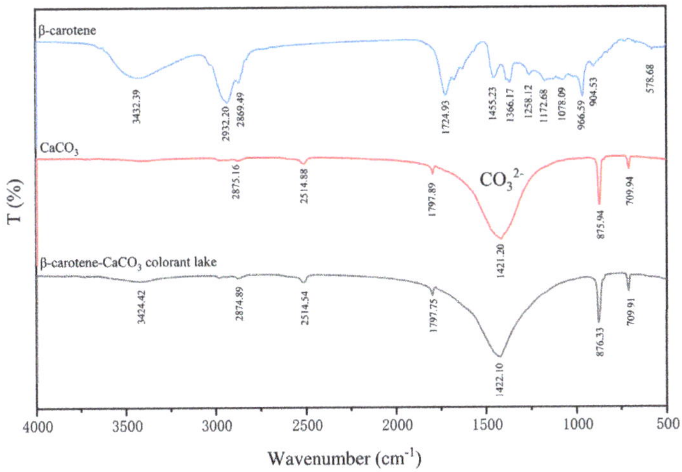


**Figure 9.** The FTIR curves of the β-carotene-$CaCO_3$ lake, $CaCO_3$, and β-carotene.

**Table 4.** The FTIR spectral peak assignments for the $CaCO_3$ and β-carotene.

| | Wavenumber ($cm^{-1}$) | Peak Assignments | References |
|---|---|---|---|
| $CaCO_3$ | 1421.20 | The asymmetric $CO_3^{2-}$ extension | [24,48] |
| | 875.94 | The $CO_3^{2-}$ plane bending vibration | |
| | 709.94 | The O-C-O group bending vibration | |
| β-carotene | 3432.39 | Stretching vibration of O-H group | [49–54] |
| | 2869.49 | Symmetric stretching vibration of C-H | |
| | 1724.93 | The C=C from the extension of the terminal cyclohexene group | |
| | 1366.17/1455.23 | The C–H from olefin | |
| | 966.59 | The C-H plane from conjugated olefin | |

## 4. Conclusions

The preparation technology of colorant lakes can improve the stability and dyeing ability of pigments, and expand their application scope in foods. *Monascus* pigment is a water-soluble pigment and β-carotene is a fat-soluble pigment. There are big differences between the two pigments in terms of structure and function. In the literature, only colorant lake preparation techniques for water-soluble pigments have been reported, and the current manuscript presents a study on the preparation of fat-soluble pigment for the first time in the literature. The results of this study confirm that coprecipitation can be successfully used for β-carotene-$CaCO_3$ lake preparation. Adsorption kinetics analysis revealed that maximum IE was attained at temperature of 40 °C and pH of 9.3. Obvious adsorption and desorption were evident during lake formation. Although ethanol, ultrasonic treatment, and drying methods influenced the lake stability and IE, they did not affect crystal morphology. The isothermal adsorption curves and EDS, XPS, TGA, DSC, and FTIR analyses demonstrated that the adsorption process was primarily governed by single-layer physical adsorption without the need for chemical reactions. Electrostatic attraction may play a predominant role in the formation of the lake. The current study demonstrates that shearing operation can significantly reduce the size of β-carotene, which is an important factor influencing the adsorption process. In future studies, more methods are expected to be investigated in order to obtain smaller sizes for β-carotene to realize better adsorption of the pigment onto $CaCO_3$, such as adding food-grade reagents (ethanol, etc.) or improving the preparation method. At the same time, our team will explore the biological activity of colorant lake products and strive for the application of colorant lake products in food.

**Author Contributions:** Conceptualization, Y.L. and D.Y.; methodology, Y.L.; investigation, Y.L., L.J. and J.C.; visualization, Y.L., L.J. and J.C.; writing—original draft preparation, Y.L.; writing—review and editing, D.Y.; data curation, Y.L.; supervision, D.Y.; project administration, D.Y. and C.W.; funding acquisition, D.Y. All authors have read and agreed to the published version of the manuscript.

**Funding:** This research was funded by the National Natural Science Foundation of China (No. 32101877).

**Institutional Review Board Statement:** Not applicable.

**Informed Consent Statement:** Not applicable.

**Data Availability Statement:** The data presented in this study are available on request from the corresponding author. The data are not publicly available due to privacy restrictions.

**Conflicts of Interest:** The authors declare that they have no known competing financial interests or personal relationships that could have appeared to influence the work reported in this paper.

## References

1. Castro-Munoz, R.; Correa-Delgado, M.; Cordova-Almeida, R.; Lara-Nava, D.; Chavez-Munoz, M.; Velasquez-Chavez, V.F.; Hernandez-Torres, C.E.; Gontarek-Castro, E.; Ahmad, M.Z. Natural sweeteners: Sources, extraction and current uses in foods and food industries. *Food Chem.* **2022**, *370*, 130991. [CrossRef] [PubMed]
2. Zin, M.M.; Marki, E.; Banvolgyi, S. Conventional extraction of betalain compounds from beetroot peels with aqueous ethanol solvent. *Acta Aliment.* **2020**, *49*, 163–169. [CrossRef]
3. Garza-Cadena, C.; Ortega-Rivera, D.M.; Machorro-García, G.; Gonzalez-Zermeño, E.M.; Homma-Dueñas, D.; Plata-Gryl, M.; Castro-Muñoz, R. A comprehensive review on Ginger (*Zingiber officinale*) as a potential source of nutraceuticals for food formulations: Towards the polishing of gingerol and other present biomolecules. *Food Chem.* **2023**, *413*, 135629. [CrossRef]
4. Shao, Y.; Tang, C. Characteristics and oxidative stability of soy protein-stabilized oil-in-water emulsions: Influence of ionic strength and heat pretreatment. *Food Hydrocoll.* **2014**, *37*, 149–158. [CrossRef]
5. Liu, Y.; Hou, Z.; Lei, F.; Chang, Y.; Gao, Y. Investigation into the bioaccessibility and microstructure changes of β-carotene emulsions during in vitro digestion. *Innov. Food Sci. Emerg. Technol.* **2012**, *15*, 86–95. [CrossRef]
6. Zhong, L.; Ma, N.; Wu, Y.; Zhao, L.; Ma, G.; Pei, F.; Hu, Q. Gastrointestinal fate and antioxidation of β-carotene emulsion prepared by oat protein isolate-*Pleurotus ostreatus* β-glucan conjugate. *Carbohydr. Polym.* **2019**, *221*, 10–20. [CrossRef] [PubMed]
7. Wang, H.; Hu, L.; Peng, L.; Du, J.; Lan, M.; Cheng, Y.; Ma, L.; Zhang, Y. Dual encapsulation of β-carotene by β-cyclodextrin and chitosan for 3D printing application. *Food Chem.* **2022**, *378*, 132088. [CrossRef] [PubMed]
8. Gomez-Mascaraque, L.G.; Perez-Masia, R.; Gonzalez-Barrio, R.; Periago, M.J.; Lopez-Rubio, A. Potential of microencapsulation through emulsion-electrospraying to improve the bioaccesibility of β-carotene. *Food Hydrocoll.* **2017**, *73*, 1–12. [CrossRef]
9. Moraes, M.; Carvalho, J.M.P.; Silva, C.R.; Cho, S.; Sola, M.R.; Pinho, S.C. Liposomes encapsulating beta-carotene produced by the proliposomes method: Characterisation and shelf life of powders and phospholipid vesicles. *Int. J. Food Sci. Technol.* **2013**, *48*, 274–282. [CrossRef]
10. Salvia-Trujillo, L.; Verkempinck, S.; Rijal, S.K.; Van Loey, A.; Grauwet, T.; Hendrickx, M. Lipid nanoparticles with fats or oils containing β-carotene: Storage stability and in vitro digestibility kinetics. *Food Chem.* **2019**, *278*, 396–405. [CrossRef] [PubMed]
11. Yildiz, Z.I.; Topuz, F.; Kilic, M.E.; Durgun, E.; Uyar, T. Encapsulation of antioxidant beta-carotene by cyclodextrin complex electrospun nanofibers: Solubilization and stabilization of beta-carotene by cyclodextrins. *Food Chem.* **2023**, *423*, 136284. [CrossRef] [PubMed]
12. Bockuviene, A.; Sereikaite, J. Preparation and characterisation of novel water-soluble β-carotene-chitooligosaccharides complexes. *Carbohydr. Polym.* **2019**, *225*, 115226. [CrossRef] [PubMed]
13. Yuan, D.; Wang, C.; Liu, Y.; Lei, S.; Wen, Y. Mechanism Study of the Formation of Indigo Carmine-Aluminum Hydroxide Lake for Food Use. *J. Food Sci.* **2019**, *84*, 3129–3139. [CrossRef]
14. Zhang, Y.; Huan, J.; Gao, D.; Xu, S.; Han, X.; Song, J.; Wang, L.; Zhang, H.; Niu, Q.; Lu, X. Blood pressure mediated the effects of cognitive function impairment related to aluminum exposure in Chinese aluminum smelting workers. *Neurotoxicology* **2022**, *91*, 269–281. [CrossRef] [PubMed]
15. Bhattacharjee, S.; Zhao, Y.; Hill, J.M.; Percy, M.E.; Lukiw, W.J. Aluminum and its potential contribution to Alzheimer's disease (AD). *Front. Aging Neurosci.* **2014**, *6*, 62. [CrossRef] [PubMed]
16. VanDuyn, N.; Settivari, R.; LeVora, J.; Zhou, S.; Unrine, J.; Nass, R. The metal transporter SMF-3/DMT-1 mediates aluminum-induced dopamine neuron degeneration. *J. Neurochem.* **2013**, *124*, 147–157. [CrossRef] [PubMed]
17. Yokel, R.A. Blood-brain barrier flux of aluminum, manganese, iron and other metals suspected to contribute to metal-induced neurodegeneration. *J. Alzheimer's Dis.* **2006**, *10*, 223–253. [CrossRef] [PubMed]
18. Tomljenovic, L. Aluminum and Alzheimer's Disease: After a Century of Controversy, Is there a Plausible Link? *J. Alzheimer's Dis.* **2011**, *23*, 567–598. [CrossRef]

19. Willhite, C.C.; Karyakina, N.A.; Yokel, R.A.; Yenugadhati, N.; Wisniewski, T.M.; Arnold, I.M.F.; Momoli, F.; Krewski, D. Systematic review of potential health risks posed by pharmaceutical, occupational and consumer exposures to metallic and nanoscale aluminum, aluminum oxides, aluminum hydroxide and its soluble salts. *Crit. Rev. Toxicol.* **2014**, *44*, 1–80. [CrossRef] [PubMed]
20. Campbell, A.; Becaria, A.; Lahiri, D.K.; Sharman, K.; Bondy, S.C. Chronic exposure to aluminum in drinking water increases inflammatory parameters selectively in the brain. *J. Neurosci. Res.* **2004**, *75*, 565–572. [CrossRef] [PubMed]
21. *GB 2760-2014*; National Food Safety Standard for Uses of Food Additives. 2014. Available online: http://down.foodmate.net/standard/sort/3/42543.html (accessed on 20 February 2024). (In Chinese)
22. *GB 2760-2011*; National Food Safety Standard for Uses of Food Additives. 2011. Available online: http://down.foodmate.net/standard/sort/3/28220.html (accessed on 20 February 2024). (In Chinese)
23. Zhang, H.; Zhang, J.Y.; Wang, H.L.; Luo, P.J.; Zhang, J.B. The Revision of Aluminum-containing Food Additive Provisions in China. *Biomed. Environ. Sci.* **2016**, *29*, 461–466. [CrossRef] [PubMed]
24. Yuan, D.; Bai, G.; Liu, Y.; Jing, L.; Wang, C.; Liu, G. A novel edible colorant lake prepared with $CaCO_3$ and Monascus pigments: Lake characterization and mechanism study. *Food Chem.* **2023**, *410*, 135408. [CrossRef] [PubMed]
25. Kim, M.; Lee, J.; Jo, M.; Kim, M.; Kim, H.; Oh, J.; Song, N.W.; Choi, S. Cytotoxicity, Uptake Behaviors, and Oral Absorption of Food Grade Calcium Carbonate Nanomaterials. *Nanomaterials* **2015**, *5*, 1938–1954. [CrossRef] [PubMed]
26. Konar, S.K.; Gu, R.; Sain, M. Preparation and characterization of baked nitrile latex foam reinforced with biomasses. *Ind. Crops Prod.* **2013**, *42*, 261–267. [CrossRef]
27. Sigurdson, G.T.; Tang, P.; Giusti, M.M. Natural Colorants: Food Colorants from Natural Sources. *Annu. Rev. Food Sci. Technol.* **2017**, *8*, 261–280. [CrossRef] [PubMed]
28. Yuan, D.; Lei, S.; Gao, L.; Liu, Y.; Liu, G.; Wang, C. Mechanism study on the formation of complex between Ca(II) and indigo carmine for potential food use as a colorant lake. *Food Chem.* **2020**, *322*, 126709. [CrossRef] [PubMed]
29. Zhang, L.; Wei, Y.; Liao, W.; Tong, Z.; Wang, Y.; Liu, J.; Gao, Y. Impact of trehalose on physicochemical stability of β-carotene high loaded microcapsules fabricated by wet-milling coupled with spray drying. *Food Hydrocoll.* **2021**, *121*, 106977. [CrossRef]
30. Da Costa, R.; Ghobril, C.; Perrin, P.; Sanson, N. Adsorption of sodium alginate on calcium carbonate microparticles: Effect of molar mass and composition. *Colloids Surf. A Physicochem. Eng. Asp.* **2024**, *682*, 132782. [CrossRef]
31. Pokrovsky, O.S.; Mielczarski, J.A.; Barres, O.; Schott, J. Surface speciation models of calcite and dolomite/aqueous solution interfaces and their spectroscopic evaluation. *Langmuir* **2000**, *16*, 2677–2688. [CrossRef]
32. Chibowski, E.; Hotysz, L.; Szcześ, A. Time dependent changes in zeta potential of freshly precipitated calcium carbonate. *Colloids Surf. A Physicochem. Eng. Asp.* **2003**, *222*, 41–54. [CrossRef]
33. Aly, Z.; Luca, V. Uranium extraction from aqueous solution using dried and pyrolyzed tea and coffee wastes. *J. Radioanal. Nucl. Chem.* **2013**, *295*, 889–900. [CrossRef]
34. Sen, T.K.; Afroze, S.; Ang, H.M. Equilibrium, Kinetics and Mechanism of Removal of Methylene Blue from Aqueous Solution by Adsorption onto Pine Cone Biomass of Pinus radiata. *Water Air Soil Pollut.* **2011**, *218*, 499–515. [CrossRef]
35. Koga, N.; Yamane, Y.; Kimura, T. Thermally induced transformations of calcium carbonate polymorphs precipitated selectively in ethanol/water solutions. *Thermochim. Acta* **2011**, *512*, 13–21. [CrossRef]
36. Yan, G.; Wang, L.; Huang, J. The crystallization behavior of calcium carbonate in ethanol/water solution containing mixed nonionic/anionic surfactants. *Powder Technol.* **2009**, *192*, 58–64. [CrossRef]
37. Zhang, L.; Liao, W.; Tong, Z.; Wei, Y.; Wang, Y.; Liu, J.; Mao, L.; Yuan, F.; Gao, Y. Modulating physicochemical properties of β-carotene in the microcapsules by polyphenols co-milling. *J. Food Eng.* **2023**, *359*, 111691. [CrossRef]
38. Lee, S.; Kim, Y.; Lee, Y.; Guim, H.; Han, S.M. Behavior and characteristics of amorphous calcium carbonate and calcite using $CaCO_3$ film synthesis. *Mater. Des.* **2016**, *112*, 367–373. [CrossRef]
39. Wang, Y.; You, Z.; Ma, K.; Dai, C.; Wang, D.; Wang, J. Corrosion resistance of a superhydrophobic calcium carbonate coating on magnesium alloy by ultrasonic cavitation-assisted chemical conversion. *Corros. Sci.* **2023**, *211*, 110841. [CrossRef]
40. Harnkarnsujarit, N.; Charoenrein, S. Influence of collapsed structure on stability of β-carotene in freeze-dried mangoes. *Food Res. Int.* **2011**, *44*, 3188–3194. [CrossRef]
41. Boyetey, M.B.; Torgbo, S.; Sukyai, P.; Watthanasakphuban, N.; Kamonsutthipaijit, N. Filter cake-derived calcium carbonate polymorphs from sugar refinery for hydroxyapatite production as a sustainable material for biomedical application. *Ceram. Int.* **2023**, *49*, 23417–23425. [CrossRef]
42. Cui, H.T.; Hong, G.Y.; You, H.P.; Wu, X.Y. Coating of $Y_2O_3$: $Eu^{3+}$ particles with alumina by a humid solid state reaction at room temperature. *J. Colloid Interface Sci.* **2002**, *252*, 184–187. [CrossRef] [PubMed]
43. Zou, Z.; Yang, X.; Alberic, M.; Heil, T.; Wang, Q.; Pokroy, B.; Politi, Y.; Bertinetti, L. Additives Control the Stability of Amorphous Calcium Carbonate via Two Different Mechanisms: Surface Adsorption versus Bulk Incorporation. *Adv. Funct. Mater.* **2020**, *30*, 2000003. [CrossRef]
44. Li, X.; Lv, Y.; Ma, B.; Wang, W.; Jian, S. Decomposition kinetic characteristics of calcium carbonate containing organic acids by TGA. *Arab. J. Chem.* **2017**, *10*, S2534–S2538. [CrossRef]
45. Wondu, E.; Lule, Z.C.; Kim, J. Polyisocyanate-based water-soluble polyurethane/$CaCO_3$ composites for gunpowder storage. *Polym. Test.* **2021**, *99*, 107211. [CrossRef]
46. Chen, B.H.; Huang, J.H. Degradation and isomerization of chlorophyll a and β-carotene as affected by various heating and illumination treatments. *Food Chem.* **1998**, *62*, 299–307. [CrossRef]

47. Xiaowan, Z.; Xunyu, L.; Huiqin, L. Degradation of β-carotene and its influence in foods. *Food Ind. Sci. Technol.* **2011**, *03*, 417–421. [CrossRef]
48. Yu, S.; Wang, X.; Wu, D. Microencapsulation of n-octadecane phase change material with calcium carbonate shell for enhancement of thermal conductivity and serving durability: Synthesis, microstructure, and performance evaluation. *Appl. Energy* **2014**, *114*, 632–643. [CrossRef]
49. Kim, W.; Wang, Y.; Vongsvivut, J.; Ye, Q.; Selomulya, C. On surface composition and stability of β-carotene microcapsules comprising pea/whey protein complexes by synchrotron-FTIR microspectroscopy. *Food Chem.* **2023**, *426*, 136565. [CrossRef]
50. Salem, A.; Ramadan, A.R.; Shoeib, T. Entrapment of β-carotene and zinc in whey protein nanoparticles using the pH cycle method: Evidence of sustained release delivery in intestinal and gastric fluids. *Food Biosci.* **2018**, *26*, 161–168. [CrossRef]
51. Siqi Qiang, J.Z.T.Y. Structure, stability and in vitro digestion of a novel zein-based oil gel delivery system loaded β-carotene. *J. Food Eng.* **2024**, *366*, 111848. [CrossRef]
52. Zarif, B.; Shabbir, S.; Shahid, R.; Noor, T.; Imran, M. Proteosomes based on milk phospholipids and proteins to enhance the stability and bioaccessibility of β-carotene. *Food Chem.* **2023**, *429*, 136841. [CrossRef] [PubMed]
53. Zhang, L.; Liao, W.; Wang, Y.; Tong, Z.; Li, Q.; Gao, Y. Thermal-induced impact on physicochemical property and bioaccessibility of β-carotene in aqueous suspensions fabricated by wet-milling approach. *Food Control* **2022**, *141*, 109155. [CrossRef]
54. Baek, E.J.; Garcia, C.; Shin, G.H.; Kim, J.T. Improvement of thermal and UV-light stability of β-carotene-loaded nanoemulsions by water-soluble chitosan coating. *Int. J. Biol. Macromol.* **2020**, *165*, 1156–1163. [CrossRef] [PubMed]

foods

Article

# Purification and Structural Characterization of Polysaccharides from *Polygonum multiflorum* Thunb. and Their Immunostimulatory Activity in RAW264.7 Cells

**Yan Gou [1], Donglin Gu [1], Jing Fan [2], Jianbo Yang [2], Ke Zan [2], Jingjing Liu [2], Hongyu Jin [2], Ying Wang [2,*], Feng Wei [2] and Shuangcheng Ma [3,*]**

1 NMPA Key Laboratory of Quality Evaluation of Chinese Patent Medicines, Sichuan Institute for Drug Control, Chengdu 611731, China; gouyan@scsyjs.org (Y.G.); 13348821019@163.com (D.G.)
2 Institute for Control of Chinese Traditional Medicine and Ethnic Medicine, National Institutes for Food and Drug Control, Beijing 102629, China; fanjing092021@163.com (J.F.); yangjianbo82@nifdc.org.cn (J.Y.); zanke@nifdc.org.cn (K.Z.); 18911002859@126.com (J.L.); jhyw@nifdc.org.cn (H.J.); weifeng@nifdc.org.cn (F.W.)
3 Chinese Pharmacopoeia Committee, Beijing 100061, China
* Correspondence: wangying17@nifdc.org.cn (Y.W.); masc@nifdc.org.cn (S.M.); Tel.: +86-10-53852076 (S.M.); Fax: +86-10-53852072 (S.M.)

**Abstract:** *Polygonum multiflorum* Thunb. (PM) and derived products are broadly utilized in Chinese traditional medicine. According to our previous research, PM mostly contains polysaccharides, which display a wide range of biological activities. Two water-soluble polysaccharides (PMPs-1 and PMPs-2) were obtained from PM by DEAE-Cellulose and Sephadex G-100 column chromatography. Colorimetry, HPGPC-MALLS-RID, HPLC-PDA, methylation, FT-IR, NMR, and SEM were used to characterize these polysaccharides. PMPs-1 and PMPs-2 had average molecular weights of 255.5 and 55.7 kDa, respectively. PMPs-1 consisted of Man, Glc, Gal, and Ara at 0.9:78.6:1.0:1.6 and was a glucan with $\rightarrow$ 4)-Glc*p*-(1 $\rightarrow$ as a backbone. Meanwhile, PMPs-2, an acidic polysaccharide, comprised Rha, GalA, Glc, Gal, and Ara at 3.2:20.3:2.7:1.0:8.3. PMPs-1 and PMPs-2 significantly improved the proliferation of RAW 264.7 cells and induced NO, TNF-$\alpha$, and IL-6 release. This study reveals that these two polysaccharides can be explored as novel immunomodulators and provide a basis for further development of PM in food and pharmaceutical industries.

**Keywords:** *Polygonum multiflorum* Thunb.; purification; polysaccharides; structural characterization; immunostimulatory

## 1. Introduction

The food market has experienced a profound and rapid transformation due to increasing interest among individuals in natural products, which harbor demonstrated health-enhancing properties. Polysaccharides, which are abundant in living organisms, represent a prevalent category of large molecular compounds with a long history of application within the food industry. Multiple reports have demonstrated diverse biological activities for plant polysaccharides, including immune regulation, antioxidant effects, hypoglycemic properties, anti-tumor activity, hypolipidemic activity, and anticoagulant effects, as well as relatively low toxicity [1–4]. These benefits indicate the significant potential of plant polysaccharides to be used in the development of healthy foods and medicinal preparations [5]. With the market demand as a driving force, researchers are actively exploring novel polysaccharides derived from animals, plants, and microorganisms for therapeutic benefits [6,7].

*Polygonum multiflorum* Thunb. (PM) root, commonly referred to as "Heshouwu" in China, has a widespread distribution across the globe. With a rich history spanning hundreds of years, PM and derived products are considered prominent herbal preparations in China and numerous East Asian countries [8]. A comprehensive analysis of the available evidence indicates that

PM can be used to alleviate aging, hyperlipidemia, cancer, and inflammation, showing great potential in the field of drug and health care product development [9]. Moreover, this botanical remedy exhibits additional benefits, including immune regulatory capabilities, neuroprotective effects, and the ability to alleviate various other ailments. Its multifaceted nature positions PM as a promising candidate for addressing complex health conditions. The fundamental constituents that contribute to the efficacy of PM primarily comprise stilbene glycosides, anthraquinones, phenols, phospholipids, and polysaccharides [10–12]. Polysaccharides make up more than 2% of PM, representing its main components. Polysaccharides regulate immunity and antioxidant activity. In a study promoted by Zhang et al., two polysaccharides, namely, WPMP-1 and WPMP-2, were purified from raw PM. Composition analysis revealed that WPMP-1 consisted exclusively of glucan (100%), while WPMP-2 exhibited a complex composition comprising arabinose (48%), rhamnose (23%), galactose (15%), and galacturonic acid (14%). To examine the potential immunomodulatory effects of these polysaccharides, researchers attempted to evaluate their effects on the proliferation of splenocytes and the phagocytic properties of peritoneal macrophages. These investigations aimed to determine whether WPMP-1 and WPMP-2 effectively activate these vital immune functions providing compelling evidence indicating the immunomodulatory potential of polysaccharides derived from PM. PM polysaccharides predominantly comprise neutral sugars, namely, glucan or arabinose, constituting a substantial majority of over 60% in total composition [13]. However, there is no systematic structural analysis and immunological activity study on the acidic polysaccharides from PM.

In this study, two water-soluble polysaccharides extracted from PM were characterized, including PMPs-1 and PMPs-2. The characterization process involved a comprehensive array of analytical techniques, such as HPLC, high-performance gel permeation chromatography/multi-angle laser light scattering/refractive index detector (HPGPC-MALLS-RID), methylation analysis, scanning electron microscopy (SEM), Fourier-transform infrared spectroscopy (FT-IR), and nuclear magnetic resonance (NMR). By systematically investigating the immunopotentiation abilities of these polysaccharides in RAW 264.7 cells, this study aimed to provide deeper insights into the intricate associations of the structural features of PMPs-1 and PMPs-2 with their pharmacological properties. The outcomes of this study provide valuable guidance for harnessing the potential of PM polysaccharides as functional food ingredients or immunomodulatory agents, promoting their enhanced utilization in various applications.

## 2. Materials and Methods

### 2.1. Materials

The PM was collected at a local market in Henan province, China. DEAE cellulose-52 and Sephadex G-100 were provided by Solarbio (Beijing, China). The National Institute for Food and Drug Control (Beijing, China) provided rhamnose (Rha), glucuronic acid (GlcA), galactose (Gal), mannose (Man), glucose (Glc), and arabinose (Ara) standards. Trifluoroacetic acid (TFA) was distributed by Oka (Beijing, China). Sodium borodeuteride ($NaBD_4$) and 1-phenyl-3-methyl-5-pyrazolone (PMP) were provided by Aladdin (Shanghai, China). The remaining reagents were of analytical grade. RAW 264.7 macrophages were from the BeNa Biotechnology Research Institute based in Beijing, China. The assay kits for NO, TNF-α, and IL-6 detection were provided by Jiangsu Enzyme-free Industrial, headquartered in Jiangsu (Nanjing, China).

### 2.2. Extraction, Isolation, and Purification of PM Polysaccharides

#### 2.2.1. Extraction of PRMPs

Polysaccharide extraction was based on water decoction and alcohol precipitation [14]. Briefly, samples were added to 80% ethanol and heated for 1 h at 85 °C. After drying, extraction was performed with water (1:20, $m/v$) for 2 h at 100 °C. The samples were centrifuged (5000 rpm for 10 min) and supernatants were concentrated by evaporation to 10 mL. Next, ethanol was added to a final concentration of 80% ($v/v$) for overnight incubation at 4 °C. The precipitates obtained by centrifugation underwent an ethanol wash. The Sevag's method was utilized to remove proteins. Small molecular compounds were removed with molecular weight retention centrifuge tubes (10 k MWCO, Thermo, Beijing, China). Finally, the precipitates were lyophilized to obtain the polysaccharide of PM (RPMPs).

#### 2.2.2. Separation and Purification of RPMPs

DEAE cellulose-52 anion exchange column chromatography was employed (2.6 × 40 cm) to separate and purify RPMPs. Elution utilized ultrapure water containing 0.3 mol/L NaCl as the eluent, at 3 mL/min, ensuring efficient separation of the desired components. To monitor the purification progress, each fraction was collected during elution, with a collection volume of 10 mL/tube. Finally, the obtained fractions were analyzed by the phenol-sulfuric acid method to detect polysaccharides.

After the initial separation, the fractions containing the target polysaccharides were concentrated by employing a rotary evaporator under reduced pressure, at 50 °C. To further refine the purification, the concentrated samples were using a Sephadex G-100 column with dimensions of 2.6 × 40 cm. This step utilized ultrapure water as the eluent, at 0.5 mL/min, aiming to separate different components based on their molecular sizes and properties. Throughout the fractionation process, 10 mL fractions were obtained and preserved for subsequent analysis. The same analytical methods mentioned above were employed to analyze the collected fractions. After concentration, the samples were dialyzed with ultrapure water at 4 °C for 48 h, and PMPs-1 and PMPs-2 were obtained by vacuum freeze-drying.

### 2.3. *Structural Characterization of Polysaccharides*

#### 2.3.1. Analytical Methods

Quantification of total sugar content was carried out by the phenol-sulfuric acid method, and glucose was employed as a reference standard for calibration [15]. Furthermore, uronic acid quantitation utilized the M-hydroxybiphenyl method, with galacturonic acid (GalA) as a reference standard for calibration [16].

#### 2.3.2. *Mw* Assessment

To determine the *Mw* of PMPs-1 and PMPs-2, HPGPC-MALLS-RID was carried out, as outlined previously [17]. During the analysis, elution was carried out by employing a mobile phase consisting of 0.1 mol/L NaCl at 0.5 mL/min.

#### 2.3.3. Monosaccharide Compositions

Derivatization of samples using PMP (1-phenyl-3-methyl-5-pyrazolone) was based on an established method [18]. In this process, PMPs-1 and PMPs-2, at a concentration of 1 mg/mL, were subjected to hydrolysis using TFA at a concentration of 4 mol/L. Hydrolysis was performed in sealed tubes under nitrogen at 120 °C for 2 h. TFA was removed by washing with methanol three times, and the hydrolysate was dissolved in ultrapure water (1 mL). Subsequently, a reaction mixture was prepared with 1 mL of the polysaccharide hydrolysate, 300 μL of a methanol solution containing the PMP at 0.5 mol/L, and 150 μL of 0.25 mol/L NaOH. The reaction was carried out under controlled conditions at 70 °C for 2 h. Finally, 150 μL of a 0.25 mol/L HCl solution was used to stop the reaction at ambient temperature.

The mixture was extracted with 1 mL chloroform and repeated three times. The supernatant was filtered through a 0.45 μm pore membrane for high-performance liquid chromatography (HPLC). The HPLC system employed was manufactured by Waters (Acquity H-class, Waters, Milford, MA, USA). HPLC conditions were as follows: ZORBAX Eclipse XDB-$C_{18}$ column (100 × 2.1 mm, particle size of 1.8 μm) provided by Agilent (Santa Clara, CA, USA); mobile phase, 84% solution A (0.1 mol/L $KH_2PO_4$ at a pH of 6.9) and 16% B (acetonitrile); detection wavelength, 250 nm; column temperature, 25 °C; flow rate, 0.3 mL/min; injection volume, 10 μL.

2.3.4. Methylation Analysis

Methylation analysis was performed using the Hakomori method [19], with slight modifications. Firstly, PMPs-2 (10 mg) was reduced using $NaBD_4$ and repeated three times. Notably, this process involved the introduction of deuterium-labeled methylene groups onto the carboxyl carbon of GalA. Consequently, GalA quantification was performed by examining the $m/z$ of Gal via GC-MS. The reduction products of PMPs-1 and PMPs-2 underwent methylation. This involved three treatment cycles with 2 mL of $CH_3I$ at ambient temperature. Next, the samples were incubated with TFA (2 mol/L) at 120 °C for 2 h. The resulting hydrolysates were then subjected to reduction by $NaBD_4$. Acetylation was carried out by incubation with pyridine and acetic anhydride. The partially methylated alditol acetates (PMAAs) were assessed with a GC–MS system (GCMS-TQ8050 NX, Shimadzu, Kyoto, Japan) equipped with a mass spectrometer and an SP-2330 capillary column (30 m × 0.25 mm, 2 μm), with a starting temperature of 80 °C, increased by 30 °C/min to 170 °C, and then by 4 °C/min to 240 °C, which was finally held for 10 min. The carrier gas He was used at 1.10 mL/min. The split ratio and inlet, ion source, and ion interface temperatures were 10 °C, 230 °C, 230 °C, and 230 °C, respectively. The injection volume was 1 μL. In mass spectrometry, the scan mode was used to collect data, with a scan ion range of $m/z$ 50~800.

2.3.5. Infrared Spectral Analysis

FT-IR spectra were acquired on a Nicolet Nagna-IR 550 spectrophotometer (Thermo Fisher Scientific, Waltham, MA, USA), spanning from 4000 to 400 $cm^{-1}$. Firstly, PMPs-1 and PMPs-2 were ground alongside spectroscopic grade potassium bromide (KBr) powder. The resulting blend was pressed to form compacted pellets with a uniform thickness of 1 mm. These pellets served as the sample for FT-IR measurements, enabling comprehensive spectral characterization.

2.3.6. SEM

PMPs-1 and PMPs-2 were analyzed with a QUANTA 250 FEI Scanning Electron Microscope (Hillsboro, OR, USA). A 5 mg polysaccharide sample was placed on a conductive carbon film with a double-sided adhesive and transferred to the ion sputtering instrument for a 40 s gold spraying. Finally, each sample was examined by SEM at 5 kV.

2.3.7. NMR Analysis

NMR spectra for PMPs-1 and PMPs-2 were generated with a high-field 600 MHz NMR spectrometer (Bruker, Billerica, MA, USA). Briefly, 50 mg polysaccharide specimens were added to 0.5 mL deuterium oxide ($D_2O$) for NMR analysis. The NMR study included recording both $^1H$ and $^{13}C$ spectra and was conducted at 30 °C. Subsequently, the acquired NMR data were subjected to comprehensive analysis using the MestReNova 6.1.0.

*2.4. Immunoregulatory Activity Test*

RAW 264.7 cells were procured from the BeNa Biotechnology Research Institute headquartered in Beijing, China. Dulbecco's modified Eagle's medium (DMEM) manufactured by Gibco (Billings, MT, USA) was utilized for cell culture after supplementation of 10% heat-inactivated fetal bovine serum (FBS). Cells were cultured with 5% $CO_2$ at 37 °C.

2.4.1. Cell Viability

To assess the potential cytotoxicity of PMPs-1 and PMPs-2 in RAW 264.7 cells, an MTT (3-(4,5-dimethyl-2-thiazolyl)-2,5-diphenyl-2H-tetrazolium bromide) kit (Abcam, Cambridge, UK) was employed. RAW 264.7 cells at 2.5 × $10^5$/mL were seeded in a 96-well plate. Subsequently, these cells were incubated at 37 °C in a humid environment with 5% $CO_2$ for an appropriate duration. Following this preincubation phase, the culture medium was aspirated and replaced by fresh media with various concentrations (0 [blank], 25, 50, and 100 μg/mL) of both PMPs-1 and PMPs-2. Each concentration was assessed in six independent replicates. After incubation for 24 h, 20 μL of MTT (5 mg/mL) was added per well for a 4 h incubation at 37 °C. Following the incubation period, an absorbance reading at 450 nm was taken using a microplate reader. Subsequently, cell viability was assessed as follows: macrophage cell viability (%) = $(A_1/A_0) \times 100\%$, where $A_0$ denotes the absorbance value obtained from the blank control group (serving as a reference baseline) and $A_1$ pertains to the absorbance value acquired exclusively from the sample, reflecting the specific response or signal exhibited by the investigated substances.

#### 2.4.2. Determination of the Levels of NO and Cytokines

NO, TNF-α, and IL-6 amounts were quantified in RAW 264.7 cells with ELISA kits as directed by the respective manufacturers. Briefly, 100 μL of a RAW 264.7 cell suspension at $10^6$ cells/mL were seeded into individual wells of a high-quality 96-well plate. Subsequently, RAW 264.7 cells underwent a 24 h treatment with varying amounts (25, 50, 100 μg/mL) of PMPs-1 and PMPs-2. For comparison, a blank control consisting of DMEM and the positive control (LPS at 1 μg/mL) were also examined. After treatment, each cell culture supernatant was collected for further analysis. NO, TNF-α, and IL-6 levels were determined by ELISA, strictly following the instructions outlined in each ELISA kit's manual.

### *2.5. Statistical Analysis*

Data are mean ± standard deviation (SD). One-way ANOVA was performed to compare groups with SPSS 20.0 based on triplicate assays. $p < 0.05$ reflected statistical significance.

## 3. Results and Discussion

### *3.1. Purification and Chemical Composition of Polysaccharides*

Crude RPMPs were isolated by DEAE cellulose-52 anion exchange column chromatography as two major peaks, i.e., RPMP-WA and RPMP-0.3A (Figure 1A), which underwent further purification with a Sephadex G-100 column, respectively (Figure 1B,C). Finally, two heteropolysaccharides were obtained, namely, PMPs-1 and PMPs-2, making up 5.6% and 21.5% of the crude polysaccharide fractions of RPMPs, respectively.

Utilizing the phenol-sulfuric acid method and employing glucose as a reference, PMPs-1 and PMPs-2 exhibited neutral sugar contents of 95.5 ± 3.55% and 19.7 ± 2.42%, respectively. Furthermore, the quantification of uronic acid content in PMPs-1 and PMPs-2 was accomplished by the M-hydroxybiphenyl method, using galacturonic acid as a reference. The results indicated that PMPs-1 had a uronic acid content of 4.7 ± 2.12%, which was substantially lower than that of PMPs-2 (79.4 ± 3.11%). Additionally, UV spectra revealed that neither polysaccharide exhibited absorption at 280 nm, thereby signifying the absence of proteins [20]. PMP-HPLC-PDA was performed to assess the monosaccharide compositions of both PMPs-1 and PMPs-2, and the results are depicted in Figure 2. Intriguingly, PMPs-1 and PMPs-2 exhibited distinct monosaccharide profiles. PMPs-1 was composed of Man (0.9), Glc (78.6), Gal (1.0), and Ara (1.6). On the other hand, PMPs-2 had an entirely different arrangement of monosaccharides, comprising Rha (3.2), GalA (20.3), Glc (2.7), Gal (1.0), and Ara (8.3) in their respective molar ratios. The complexity of these monosaccharides suggests that PMPs-1 and PMPs-2 differ in terms of underlying molecular structure.

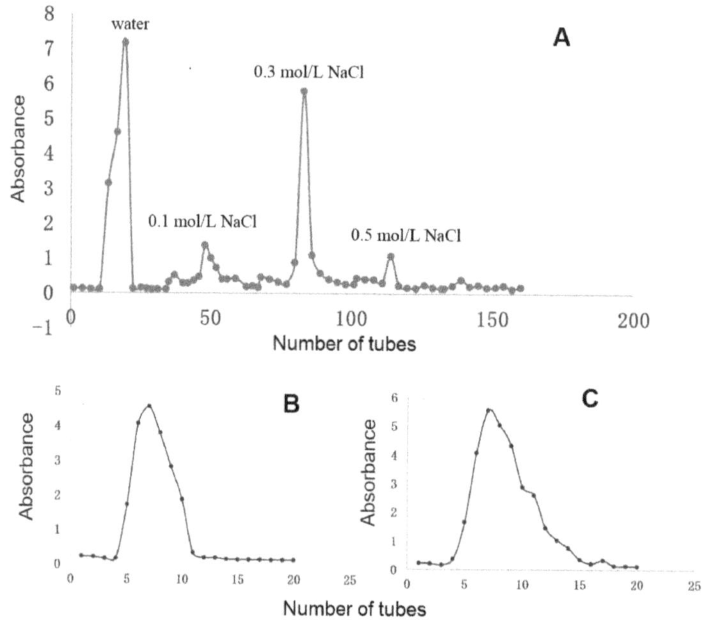


**Figure 1.** Isolation and purification of PMPs-1 and PMPs-2 from *Polygonum multiflorum* Thunb. The elution profile of RPMPs on a DEAE cellulose-52 chromatography column (**A**). The elution profiles of RPMP-WA (**B**) and RPMP-0.3A (**C**) on a Sephadex G-100 chromatography column.

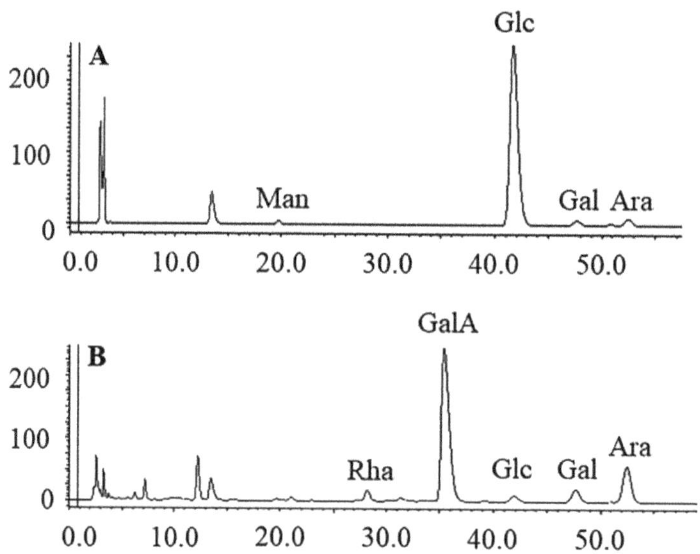


**Figure 2.** HPLC analysis of PMP derivatives of PMPs-1 (**A**) and PMPs-2 (**B**).

### 3.2. Molecular Properties of PMPs-1 and PMPs-2

The HPGPC chromatograms are shown in Figure 3, where both polysaccharides exhibited a symmetrical peak. However, the *Mw* values of PMPs-1 and PMPs-2 were different. PMPs-1 showed an *Mw* of 255.5 kDa, which was higher than the 55.7 kDa determined for PMPs-2. It is worth noting that the *Mw* results of this study differ somewhat from previous findings. For instance, Lv et al. reported *Mw* values of 480 and 610 kDa, respectively, for two purified polysaccharide components [21]. Meanwhile, *Mw* values of 2.04 and 92.13 kDa were reported for two purified polysaccharides by Zhang et al., respectively [13]. The *Mw* measurement results for PM polysaccharides were inconsistent, which may be related to divergent separation and *Mw* determination methods. In the determination of *Mw* of polysaccharides by using GPC-RID, well-characterized polysaccharide standards, such as commercially available pullulans or dextrans, should be used in the GPC column calibration. In studies by Lv et al. and Zhang et al., the GPC-RID method with polysaccharide standards was utilized to obtain relative *Mw* measurements. On the other hand, the present study adopted the HPGPC-MALLS-RID approach, which is a method for determining absolute *Mw* without relying on reference polysaccharide materials [22,23].

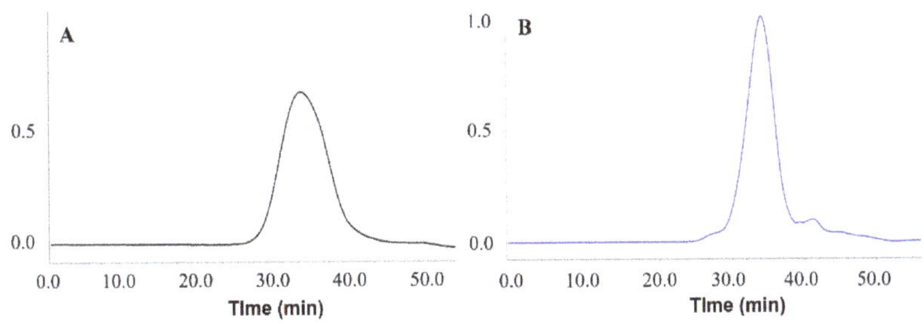


**Figure 3.** (**A**) *Mw* analysis of PMPs-1 using HPGPC-MALLS-RID; (**B**) *Mw* analysis of PMPs-2 using HPGPC-MALLS-RID.

The evaluation of polysaccharide uniformity can be accomplished by analyzing their *PDI* (*Mw*/*Mn*) values, where a value close to 1 indicates a high degree of homogeneity. In the case of PMPs-1 and PMPs-2, both showed *PDI* values of approximately 1.21 and 1.14, respectively. These values indicate that the *Mw* distribution of PMPs-1 and PMPs-2, along with their derivatives, exhibited a remarkable level of consistency and uniformity. The proximity of these *PDI* values to 1 reinforces the notion that the polysaccharides in question possess homogeneous *Mw* values without significant variations or deviations. Such findings shed light on the intrinsic stability and regularity of the polysaccharide structures under investigation.

### 3.3. Methylation Analysis of PMPs-1 and PMPs-2

The glycosidic linkages and branching of the polysaccharides were obtained by traditional methylation analysis. PMPs-1 displayed four peaks in the total ion chromatogram, indicating the presence of specific glycosidic linkages. On the other hand, PMPs-2 exhibited a more complex profile with eight distinct peaks, suggesting a broader range of glycosidic linkages within its structure compared with PMPs-1. Considering various analytical indexes, including the relative retention time and peak fragments (Supplementary File S1), and using the comprehensive Complex Carbohydrate Research Center (CCRC) Spectral Database for partially methylated alditol acetate (PMAA), the glycosidic linkages of PMPs-1 and PMPs-2 were determined. The results are summarized in Table 1, providing a comprehensive and systematic overview of the specific glycosidic linkages in each polysaccharide.

**Table 1.** Methylation analysis of PMPs-1 and PMPs-2.

| Sample | Retention Time (min) | Linkage Pattern | Mass Fragments (*m/z*) | Peak Area Ratio (%) |
|---|---|---|---|---|
| PMPs-1 | 13.840 | T-Glc*p* | 87, 101, 102, 118, 129, 145, 161, 162, 205 | 8.03 |
| | 18.014 | 1,6-Linked-Glc*p* | 102, 113, 118, 129, 173, 233 | 2.81 |
| | 18.295 | 1,4-Linked-Glc*p* | 87, 99, 102, 113, 118, 129, 173, 233 | 80.54 |
| | 21.707 | 1,4,6-Linked-Glc*p* | 118, 261 | 8.62 |
| PMPs-2 | 10.560 | T-Rha*p* | 89, 102, 118, 131, 162, 175 | 1.62 |
| | 10.950 | T-Ara*f* | 87, 102, 118, 129, 145, 161, 162 | 5.41 |
| | 13.485 | 1,2-Linked-Rha*p* | 89, 100, 115, 130, 131, 190 | 3.06 |
| | 13.830 | 1,3-Linked-Ara*f* | 87, 101, 102, 118, 129, 145, 161, 162, 205 | 5.30 |
| | 14.680 | T-Gal*p*A | 87, 102, 118, 129, 161, 162, 205 | 4.89 |
| | 15.125 | 1,5-Linked-Ara*f* | 87, 102, 118, 129, 189 | 3.38 |
| | 17.752 | 1,3,5-Linked-Ara*f* | 85, 99, 118, 127 | 5.47 |
| | 17.785 | 1,4-Linked-Gal*p*(A) | 87, 99, 102, 113, 118, 131, 175 | 63.78 |
| | 18.285 | 1,4-Linked-Glc*p* | 87, 99, 102, 113, 118, 129, 233 | 7.29 |

Results showed that the linkage types in PMPs-1 were 1,4-linked-Glc*p*, 1,6-linked-Glc*p*, 1,4,6-linked-Glc*p*, and T-linked-Glc*p*, at ratios of 80.54:2.81:8.62:8.03. These results suggested that PMPs-1 had a backbone composed of $\rightarrow$ 4)-Glc*p*-(1 $\rightarrow$ units, which was consistent with previous findings [13]. Notably, the analysis of the polysaccharide structure suggested a ratio between terminal units and branching points of 1:1.

There are more GalA units in PMPs-2; due to the poor solubility of acidic polysaccharides in methylation analysis, many hydroxyl groups could not participate in the methylation reaction, resulting in a reduced degree of methylation, and could not accurately determine the linkage sites of acidic sugars. Many hydroxyl groups could not participate in the methylation reaction, resulting in a reduced degree of methylation and the inability to accurately determine the linkage sites of acidic sugars. Therefore, GalA was esterified and reduced to Gal using the carbodiimide reagent, and the linkage sites of acidic sugars were inferred from the linkage sites analyzed by methylation after esterification and reduction. GC-MS analysis of PMPs-2 revealed at least nine easily identifiable methylated glycoside residues, including 1,2-linked-Rha*p* (3.06%), 1,3-linked-Ara*f* (5.30%), 1,5-linked-Ara*f* (3.38%), 1,3,5-linked-Ara*f* (5.47), 1,4-linked-Gal*p*A (63.78%), 1,4-linked-Glc (7.29%), T-Rha*p* (1.62%), T-Ara*f* (5.41%), and T-Gal*p*A (4.89%). The amount of Gal*p*A residues was calculated from the amount of increased galactosyl residues in the reduced polysaccharide compared with those in the native form. This result was consistent with the monosaccharide composition results. It is worth noting that Ara in PMPs-1 and PMPs-2 both existed as terminally linked arabinofuranosyl residues.

### 3.4. FT-IR Spectrum Analysis of PMPs-1 and PMPs-2

The analysis of polysaccharide structures can be effectively accomplished through infrared absorption, which arises from molecular dipole moments or charge distribution induced by vibrational motions. This technique can yield valuable information regarding the structural properties of polysaccharides. Specifically, by examining characteristic absorption peaks in the infrared spectra of these compounds, the potential structural characteristics of individual components can be determined [24]. Figure 4 shows the infrared spectra of PMPs-1 and PMPs-2, providing a comprehensive visual representation of molecular vibrations and highlighting distinct features associated with their unique structural compositions.

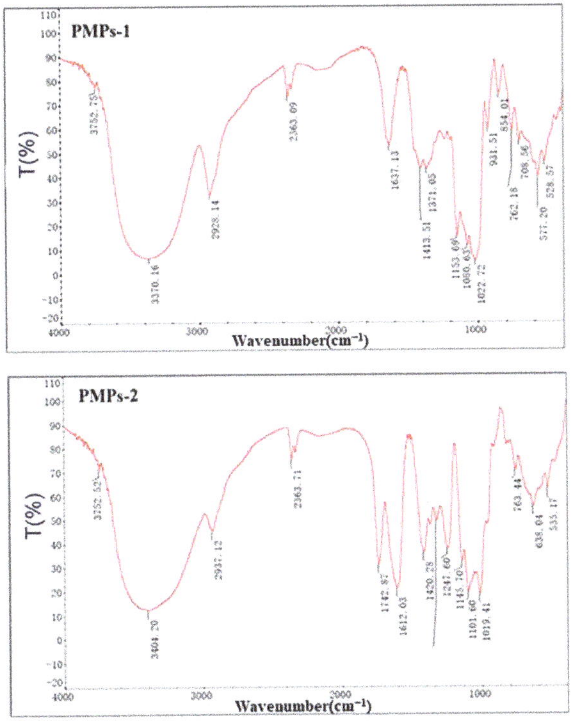


**Figure 4.** FT-IR spectra of PMPs-1 and PMPs-2.

The FT-IR spectrum of PMPs-1 revealed multiple signals spanning from 4000 to 400 $cm^{-1}$, which are in line with the expected characteristics of polysaccharides. Notably, a strong and wide stretching peak at 3370 $cm^{-1}$ was attributed to the vibrational motion of O-H groups, reflecting hydroxyl groups inherent to polysaccharides. Additionally, a relatively weaker stretching peak at 2928 $cm^{-1}$ was caused by stretching vibrations of C-H bonds within the polysaccharide molecule [25]. Bands in the region around 1419 $cm^{-1}$ corresponded to stretching vibrations due to carbon–oxygen (C-O) bonds [26]. Furthermore, a peak at 1022 $cm^{-1}$ was attributed to angular vibrations involving oxygen–hydrogen (O-H) bonds. Another peak at 854 $cm^{-1}$ is indicative of α-glycosidic linkages, which are important structural components [27]. Furthermore, an absorption peak at 1022 $cm^{-1}$ suggests the presence of angular vibrations involving oxygen–hydrogen (O-H) bonds. Another absorption peak at 854 $cm^{-1}$ is indicative of the presence of α-type glycosidic linkages, which are important structural components [28]. Lastly, a peak at 762 $cm^{-1}$ may be associated with the symmetric ring stretching vibrations of the pyran ring, a characteristic feature of the molecular structure under investigation.

The FT-IR spectrum of PMPs-2 is depicted in Figure 4. The strong absorption peak of PMPs-2 at 3404 $cm^{-1}$ was attributed to OH stretching vibrations due to hydrogen bonds of glycopyranose hydroxyl groups. Absorption peaks at 1742 $cm^{-1}$ and 1612 $cm^{-1}$ were attributed to the C=O stretching vibrations of the esterified carboxyl group (COO-R) and the asymmetric stretching vibrations of the carboxylic acid anion (COO-), respectively, suggesting uronic acid was included in the structure [29]. The peak at 1420 $cm^{-1}$ may be caused by C-O stretching vibrations. The peak at 1329 $cm^{-1}$ may result from the symmetric stretching vibrations of C=O. Finally, the peak at 1019 $cm^{-1}$ may be attributed to O-H deformation vibrations.

### 3.5. SEM Analysis of PMPs-1 and PMPs-2

The inherent complexity of polysaccharide structures gives rise to a multitude of microstructural variations. To gain insights into the surface morphology of polysaccharides, SEM has emerged as an investigational tool. Here, SEM was employed to assess the surface properties of PMPs-1 and PMPs-2 at different magnifications, including 1000- and 5000-fold, as depicted in Figure 5A–D. The SEM images captured for PMPs-1 and PMPs-2 offered visual information regarding their surface morphology. Notably, the images revealed that both samples predominantly consist of freely dispersed fragments rather than exhibiting a cohesive or organized structure. SEM images in Figure 5, specifically Figure 5A,B, depict a visual representation of PMPs-1. These images reveal that the surface of PMPs-1 exhibits both irregularity and smoothness, with a honeycomb-like structure. On the other hand, Figure 5C,D present SEM images for the acidic polysaccharide PMPs-2. Upon observation of these images, it is apparent that PMPs-2 is composed of layered structures with irregular shapes and relatively rough surfaces.

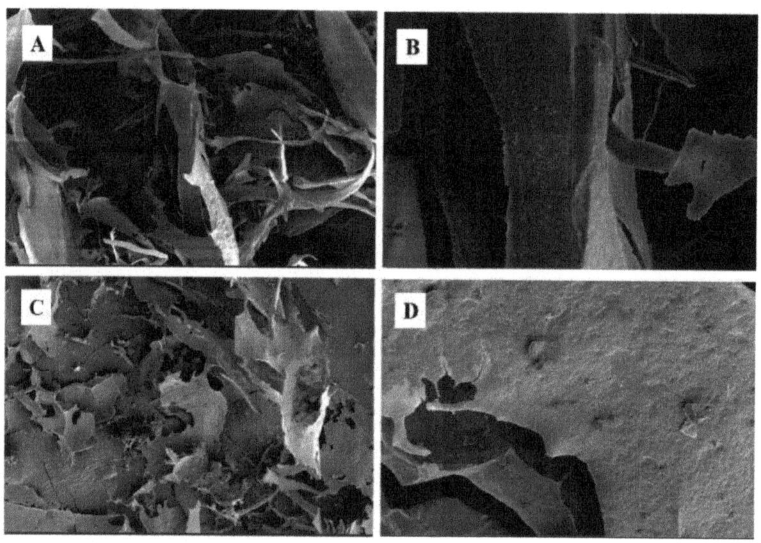


**Figure 5.** SEM images of PMPs-1 ((**A**) ×1000, (**B**) ×5000) and PMPs-2 ((**C**) ×1000, (**D**) ×5000).

### 3.6. NMR Analysis of PMPs-1 and PMPs-2

NMR analysis was employed to obtain more information about the structure of PMPs-1 and PMPs-2. Figure 6 shows the $^{1}$H and $^{13}$C NMR spectra of PMPs-1 and PMPs-2. Most of the proton signals were found at δ H 3.0~5.5 ppm and δ C 50~110 ppm, which were the typical NMR spectrums of polysaccharides. The number of anomeric proton signals at δ H 4.5~5.5 ppm and anomeric carbon signals at δ C 95~110 ppm reflects the types of monosaccharides.

Signals at 5.60~4.90 and 4.90~4.30 ppm generally correspond to the anomeric protons of α- and β-anomers, respectively. The $^{1}$H NMR spectra of PMPs-1 in $D_2O$ are depicted in Figure 6A. Most of the proton signals appeared in the region of δ 3.1–5.4 ppm, and there were three peaks with strong intensity in the anomeric region (δ 4.5–5.9 ppm) with the chemical shifts of 5.30, 5.28, and 4.87 ppm, respectively. The results indicated that PMPs-1 contained α-glycosidic linkage, which is consistent with the results of FT-IR analysis and methylation GC-MS. The $^{1}$H NMR spectra of PMPs-2 in $D_2O$ are shown in Figure 6C. Anomeric signals observed at δ 4.88, 4.86 ppm and δ 5.72, 5.32, 5.17, 5.10, 5.07, and 5.01 ppm indicated that PMPs-2 contained α- and β-glycosidic linkages. Carbohydrate ring protons have signals mostly at 3.0~4.0 ppm because of the shielding effect of the OH group [30]. The $^{13}C$ NMR spectra of PMPs-1 and PMPs-2 in $D_6O$ are shown in Figure 6B,D. Three signals appeared in the heterotopic carbon signal region of PMPs-1 (δ 99.97 ppm), versus eight signals in PMPs-2 (δ 107.80, 107.20, and 100.48 ppm). Four signals (δ 175.31, 171.15, 171.10, and 170.99 ppm) appeared in the $^{13}C$ NMR spectrum of PMPs-2, which are characteristic of the carboxyl group of uronic acid. Carbon signals between δ 60.73 and δ 77.10 ppm reflect absorption signals for $C_2$–$C_6$ in monosaccharides in PMPs-1, versus signals between δ 53.19 and δ 84.22 ppm reflect absorption signals for $C_2$–$C_6$ in PMPs-1. The NMR analysis corroborates monosaccharide composition and infrared analysis.

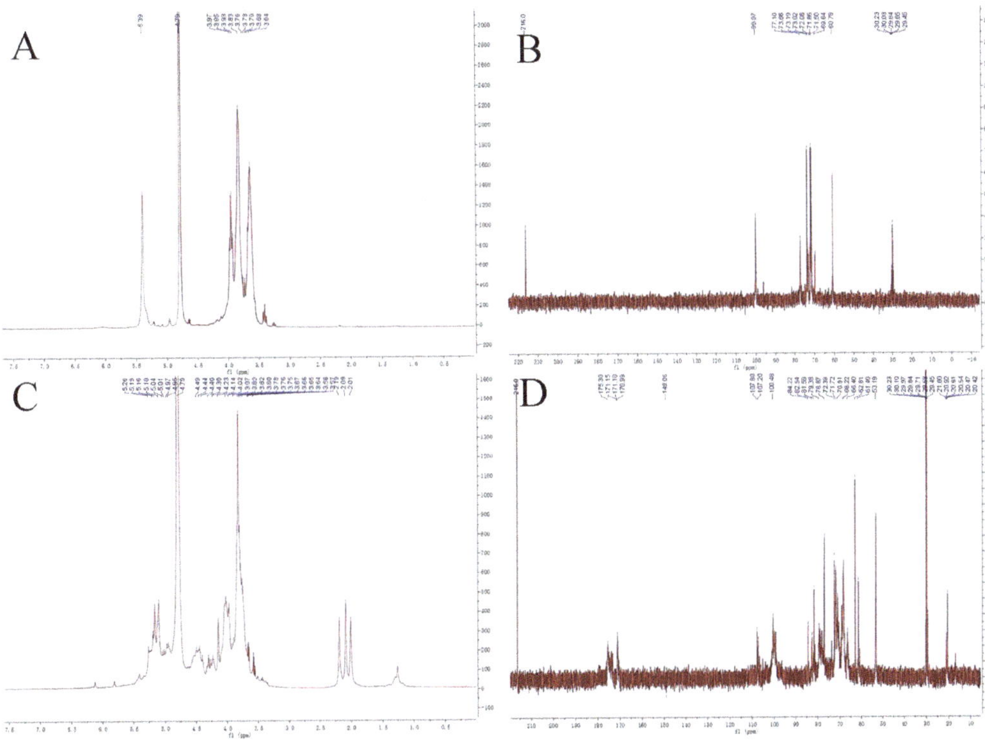


**Figure 6.** $^1H$−NMR spectra (**A**,**C**) and $^{13}C$−NMR spectra (**B**,**D**) of the purified polysaccharides PMPs-1 and PMPs-2.

### 3.7. Immunomodulatory Activities of PMPs-1 and PMPs-2

#### 3.7.1. Effects of PMPs-1 and PMPs-2 on Cell Viability

As cells trigger innate immunity against infections and inflammation, macrophages constitute a great cellular model for assessing the body's capacity to regulate the immune system. Their significance lies in their ability to serve as key players in immune responses, enabling a deeper understanding of intricate mechanisms underlying immune regulation within the body [31]. Polysaccharides typically exert their immune regulatory activities by affecting macrophages. RAW 264.7 macrophages, the prevailing model for investigating immunomodulatory functions, have been extensively utilized in previous research. Notably, polysaccharides from Atractylodis Macrocephalae Rhizoma [32], Radix Adenophorae [33], and Rehmanniae Radix Praeparata [34] exhibit potent immunoregulatory activity by promoting cell proliferation of RAW 264.7 cells. The current research aimed to investigate the impacts of PMPs-1 and PMPs-2 on cell proliferation of RAW 264.7. Remarkably, versus the blank control group, PMPs-1 and PMPs-2 exerted negligible toxic effects. Moreover, at 25–100 µg/mL ($p < 0.05$), both PMPs-1 and PMPs-2 enhanced macrophage proliferation.

These findings, visually represented in Figure 7A, underscore the stimulatory effects of PMPs-1 and PMPs-2 on macrophages.

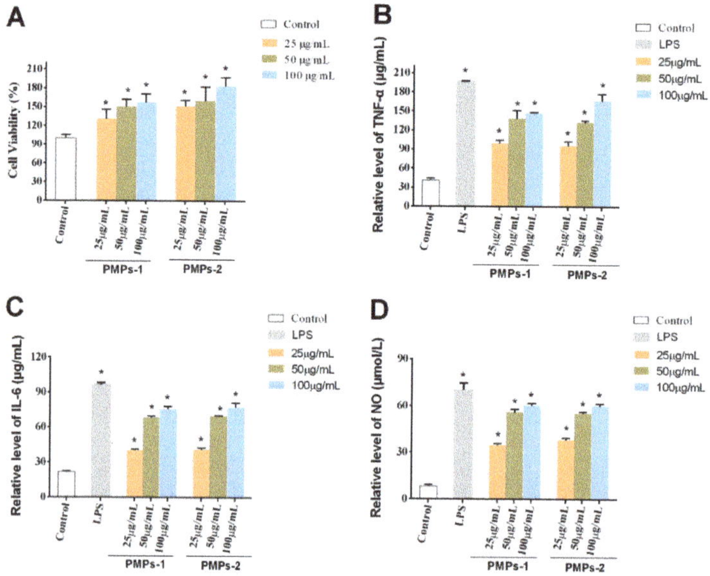

**Figure 7.** Effects of PMPs-1 and PMPs-2 on the viability (**A**), production of TNF-α (**B**), production of IL-6 (**C**), and production of NO (**D**) in RAW264.7 cells. * $p < 0.05$ denotes a statistically significant difference between the treated and control groups.

3.7.2. Effects of PMPs-1 and PMPs-2 on the Contents of Cytokines in RAW264.7 Cells

Cytokines, encompassing diverse bioactive proteins, are synthesized through the collaborative effort of immune and non-immune cells [35]. Among these cytokines, notable examples include NO, IL-6, and TNF-α [36]. Functioning as messenger molecules and cytokines play pivotal roles in multiple pathophysiological events associated with the immune response [37,38]. NO assumes a central position as a key mediator within the intricate network of macrophages. Its involvement spans essential immune-related functions and is indispensable for the immune system's ability to combat invading pathogens [39]. In the context of tissue damage from bacterial, fungal, or tumor cell assaults, macrophages exhibit a remarkable defense mechanism involving the release of large amounts of NO [40]. This immune response constitutes an important line of defense against pathogenic organisms. Intriguingly, in Figure 7B, it is evident that NO levels displayed a significant and statistically discernible difference at concentrations of 25–100 μg/mL for PMPs-1 and PMPs-2 versus control values ($p < 0.05$). TNF-α, considered a pleiotropic cytokine, plays a pivotal role in orchestrating inflammation and eliciting immune responses [41]. Within the complex framework of immune processes, IL-6 plays a vital role. This multifunctional cytokine is highly involved in various immunological activities, including phagocytosis, antigen presentation, and the fine-tuning of inflammatory responses [42]. A plethora of studies have conclusively exhibited the ability of polysaccharides to exert regulatory effects on cytokine and chemokine production. For example, a novel acidic polysaccharide named SSPA5-1 was isolated from *Scapharca subcrenata* in Li's study. The immunoregulatory activity of SSPA50-1 was evaluated on RAW 264.7 cells, and the results showed that SSPA50-1 possessed potent immunoregulatory activity by enhancing the NO, iNOS, TNF-T, and IL-6 secretion capacity of RAW 264.7 cells [43]. Sun et al. studied the immunomodulatory activity of polysaccharides from *Helicteres angustifolia* L. on RAW 264.7 cells. They found that SPF3-1 can stimulate the NO and immunomodulatory cytokines generation of RAW 264.7 cells, such as TNF-α, IL-2, IL-4, IL-6, and IL-10 [44]. Notably, these polysaccharides stimulate the release of a wide spectrum of cytokines, encompassing both anti-inflammatory and pro-inflammatory groups. Consequently, these dynamic molecular entities play essential roles in orchestrating the delicate balance of inflammatory responses within the complex network of immune processes [45,46]. The results are shown in Figure 7C,D. Notably, RAW 264.7 cells administered PMPs-1 and PMPs-2 at 25 to 100 μg/mL, which showed remarkably upregulated TNF-α and IL-6 ($p < 0.01$).

The complex associations of the immune activities of polysaccharides and their structural properties are widely recognized in the scientific community, encompassing factors such as molecular size, sulfation levels, carboxyl groups, and molecular conformation. These structural characteristics control the immunomodulatory effects of polysaccharides [47,48]. For instance, when examining the pro-inflammatory response elicited in macrophages by polysaccharides from Cystoseira indica, a close association was found between low *Mw*, sulfation content, and activity [49]. Cui et al. isolated and purified three water-soluble polysaccharides (AMAP-1, AMAP-2, and AMAP-3) from Atractylodis Macrocephalae Rhizoma. The primary immunomodulatory activity of three polysaccharides in RAW 264.7 macrophages was investigated. The results showed that AMAP-1 and AMAP-2 with high Mw can stimulate RAW 264.7 macrophages to release NO, but low *Mw* AMAP-3 rich in homogalacturonan cannot [32]. In this study, the regulatory effects of PMPs-1 and PMPs-2 on immune function were examined. The findings revealed that both PMPs-1 and PMPs-2 exerted positive effects on the immune system by stimulating TNF-α, IL-6, and NO secretion in RAW 264.7 cells. Notably, these stimulatory effects were dose-dependent at 25 to 100 μg/mL. Importantly, PMPs-1 and PMPs-2 had no significant differences in their regulatory abilities in this investigation of RAW 264.7 cells. When studying the immunomodulatory effects of plant polysaccharides with different structures using in vitro cell models, there may not be significant differences in their activity. In Zhou's study, two polysaccharides (SDH-WA and SDH-0.2A) from Rehmanniae Radix Praeparata were extracted and purified. The immunomodulatory effects of two polysaccharides were investigated by using RAW 264.7 cells. The results showed that both polysaccharides showed significantly promoted phagocytic activity, but there were no obvious differences between neutral polysaccharide SDH-WA and acidic polysaccharide SDH-0.2A. This research result is similar to our research findings. However, in vitro data do not always translate into in vivo findings [34]. In Zhang's study [13], two purified polysaccharides from PM, namely, WPMP-1 and WPMP-2, exhibited activating effects on splenocytes and macrophages. However, the acid polysaccharide WPMP-2 exhibited better immunomodulatory activity than the neutral polysaccharide WPMP-1, which indicates that the positive effects of polysaccharides may be associated with a higher content of uronic acid. Therefore, further investigation is required in the future to unveil the associated mechanism and structure–activity relationship for these polysaccharides.

## 4. Conclusions

In summary, two polysaccharides (PMPs-1 and PMPs-2) were obtained from the traditional Chinese medicine herb *Polygonum multiflorum* Thunb. The structural characterization was determined by different methods, such as UV, HPLC-PDA, SEM, FT-IR, GC-MS, HPGPC, and NMR. These two polysaccharides possessed different monosaccharide compositions and *Mw*; PMPs-1 is a neutral heteropolysaccharide, which is composed of Man, Glc, Gal, and Ara at molar ratios of 0.9, 78.6, 1.0, and 1.6, and with *Mw* of 255.5 kDa. The PMPs-2 is an acidic heteropolysaccharide, which is composed of Rha, GalA, Glc, Gal, and Ara with molar ratios of 1.6, 20.3, 2.7, 1.0, and 6.3, and with *Mw* of 55.7 kDa. In previous studies on the activity of PM polysaccharides, the focus was mainly on antioxidant activity analysis, with less research involving immunomodulatory effects. In the existing analysis of the immune activity of PM polysaccharides, in vitro and in vivo immunosuppressive models are mainly used. For example, Zhang et al. used a 5-Fu-induced macrophage immunosuppressive model and Chen et al. [50] used a cyclophosphamide-induced mice immunosuppressive model to reveal the immunomodulatory effects of PM polysaccharides. In our study, the RAW 264.7 macrophage model, a widely used model for studying the immune activity of polysaccharides, was used for studying the immunomodulatory effects of PM. The immunomodulatory activity assay showed that PMPs-1 and PMPs-2 (25~100 μg/mL) markedly enhance cell proliferation in RAW 264.7 cells. In addition, PMPs-1 and PMPs-2 increase the phagocytic activity of macrophages and starkly promote the secretion of NO and cytokines (TNF-α and IL-6) dose-dependently. In summary, these two polysaccharides can be explored as a new food additive and immune enhancer. These findings also provide a basis for further application of PM in food and pharmaceutical industries.

**Supplementary Materials:** The following supporting information can be downloaded at https://www.mdpi.com/article/10.3390/foods13060932/s1, Supplementary File S1: The mass spectrums of methylation analysis of each linkage.

**Author Contributions:** Y.G.: conceptualization, writing—review, and funding acquisition; Y.W. and D.G.: investigation, methodology, and writing—original draft; J.F.: investigation and data curation; J.Y.: investigation and methodology; K.Z.: investigation; F.W.: writing—review and editing; J.L.: writing—review and editing; H.J.: writing—review and project administration; S.M.: writing—review and editing and project administration. All authors have read and agreed to the published version of the manuscript.

**Funding:** This research was funded by the Training Fund for academic leaders of NIFDC (Grant No. 2023X8) and the open project grant of Sichuan Provincial Institute for Drug Control/NMAP Key Laboratory of Quality Evaluation of Chinese Patent Medicine (Grant Chinese Patent Medicine—2023-KFKT-001).

**Institutional Review Board Statement:** Not applicable.

**Informed Consent Statement:** Not applicable.

**Data Availability Statement:** The original contributions presented in the study are included in the article/Supplementary Material, further inquiries can be directed to the corresponding authors.

**Acknowledgments:** This work was supported by the Training Fund for academic leaders of NIFDC (Grant No. 2023X8) and the Open Project of NMPA Key Laboratory of Quality Evaluation of Chinese Patent Medicines (Sichuan Testing Center of Medical Devices/Sichuan Institute for Drug Control) (Grant No. 2023-KFKT-001).

**Conflicts of Interest:** The authors declare no conflicts of interest.

## References


1. Liu, A.J.; Yu, J.; Ji, H.Y.; Zhang, H.C.; Zhang, Y.; Liu, H.P. Extraction of a novel cold-watersoluble polysaccharide from *Astragalus membranaceus* and its antitumor and kmmunological activities. *Molecules* **2017**, *23*, 62. [CrossRef] [PubMed]
2. Souza, R.O.; Assreuy, A.M.; Madeira, J.C.; Chagas, F.D.; Parreiras, L.A.; Santos, G.R.; Mourao, P.A.; Pereira, M.G. Purified polysaccharides of *Geoffroea spinosa* barks have anticoagulant and antithrombotic activities devoid of hemorrhagic risks. *Carbohydr. Polym.* **2015**, *124*, 208–215. [CrossRef] [PubMed]
3. Tabarsa, M.; Anvari, M.; Joyner, H.S.; Behnam, S.; Tabarsa, A. Rheological behavior and antioxidant activity of a highly acidic gum from *Althaea officinalis* flower. *Food Hydrocoll.* **2017**, *69*, 432–439. [CrossRef]
4. Nie, Y.; Luo, F.; Wang, L.; Yang, T.; Shi, L.; Li, X.; Shen, J.; Xu, W.; Guo, T.; Lin, Q. Antihyperlipidemic effect of rice bran polysaccharide and its potential mechanism in high-fat diet mice. *Food Funct.* **2017**, *8*, 4028–4041. [CrossRef] [PubMed]
5. Zhao, Y.; Yan, B.C.; Wang, Z.Y.; Li, M.J.; Zhao, W. Natural polysaccharides with immunomodulatory activities. *Mini-Rev. Med. Chem.* **2020**, *20*, 96–106. [CrossRef]
6. Niu, Y.G. Introduction to the Special Issue: Preparation, Physicochemical Properties and Application of Natural Plant Polysaccharides. *Foods* **2023**, *12*, 2457. [CrossRef]
7. Ullah, S.; Khalil, A.A.; Shaukat, F.; Sources, Y.S. Extraction and Biomedical Properties of Polysaccharides. *Foods* **2019**, *8*, 304. [CrossRef]
8. Rao, T.; Liu, Y.T.; Zeng, X.C.; Li, C.P.; Ou-Yang, D.S. The hepatotoxicity of *Polygonum multiflorum*: The emerging role of the immune-mediated liver injury. *Acta. Pharmacol. Sin.* **2020**, *42*, 27–35. [CrossRef]
9. Lin, L.F.; Ni, B.R.; Lin, H.M.; Zhang, M.; Li, X.C.; Yin, X.B.; Qu, C.J.; Ni, J. Traditional usages, botany, phytochemistry, pharmacology and toxicology of *Polygonum multiflorum* Thunb.: A review. *J. Ethnopharmacol.* **2015**, *159*, 158–183. [CrossRef]

10. Liu, Y.; Wang, Q.; Yang, J.B.; Guo, X.H.; Liu, W.X.; Ma, S.C.; Li, S.P. *Polygonum multiflorum* Thunb.: A Review on Chemical Analysis, Processing Mechanism, Quality Evaluation, and Hepatotoxicity. *Front. Pharmacol.* **2018**, *9*, 364. [CrossRef]
11. Teka, T.; Wang, L.M.; Gao, J.; Mou, J.J.; Pan, G.X.; Yu, H.Y.; Gao, X.M.; Han, L.F. *Polygonum multiflorum*: Recent updates on newly isolated compounds, potential hepatotoxic compounds and their mechanisms. *J. Ethnopharmacol.* **2021**, *271*, 113864. [CrossRef]
12. Yang, J.B.; Wang, Q.; Gao, H.Y.; Wang, X.T.; Song, Y.F.; Wang, Y.; Cheng, X.L.; Wei, F.; Jin, H.T.; Ma, S.C.; et al. Dianthrones of Polygoni multiflori radix and *Polygoni multiflora* caulis: A review. *Mod. Chin. Med.* **2022**, *24*, 1431–1436.
13. Zhang, Q.; Xu, Y.; Lv, J.; Cheng, M.; Wu, Y.; Cao, K.; Zhang, X.F.; Mou, X.; Fan, Q. Structure characterization of two functional polysaccharides from *Polygonum multiflorum* and its immunomodulatory. *Int. J. Biol. Macromol.* **2018**, *113*, 195–204. [CrossRef]
14. Gu, D.L.; Wang, Y.; Jin, H.Y.; Kang, S.; Liu, Y.; Zan, K.; Fan, J.; Wei, F.; Ma, S.C. Changes of Physicochemical Properties and Immunomodulatory Activity of Polysaccharides during Processing of *Polygonum multiflorum* Thunb. *Front Pharmacol.* **2020**, *13*, 934710. [CrossRef] [PubMed]
15. Michel, D.B.; Gilles, K.A.; Hamilton, J.K.; Rebers, P.A.; Smith, F. Colorimetric Method for Determination of Sugars and Related Substances. *Anal. Chem.* **1956**, *28*, 350–356. [CrossRef]
16. Blumenkrantz, N.; AsboeHansen, G. New method for quantitative determination of uronic acids. *Anal. Biochem.* **1973**, *54*, 484–489. [CrossRef]
17. Wang, Y.; Yang, J.B.; Jin, H.Y.; Gu, D.L.; Wang, Q.; Liu, Y.; Zan, K.; Fan, J.; Wei, F.; Ma, S.C.; et al. Comparisons of physicochemical features and hepatoprotective potentials of unprocessed and processed polysaccharides from *Polygonum multiflorum* Thunb. *Int. J. Biol. Macromol.* **2023**, *235*, 123901. [CrossRef] [PubMed]
18. Wang, Y.; Jin, H.Y.; Dong, X.X.; Yang, S.; Ma, S.C.; Ni, J. Quality evaluation of *Lycium barbarum* (wolfberry) from different regions in China based on polysaccharide structure, yield and bioactivities. *Chin. Med.* **2019**, *14*, 49. [CrossRef]
19. Lee, S.J.; In, G.; Han, S.T.; Lee, M.H.; Lee, J.W.; Shin, K.S. Structural characteristics of a red ginseng acidic polysaccharide rhamnogalacturonan I with immunostimulating activity from red ginseng. *J. Ginseng Res.* **2020**, *44*, 570–579. [CrossRef]
20. Liu, C.Y.; Hu, D.J.; Zhu, H.; Zhang, Y.Y.; Qin, J.; Wang, F.; Zhang, Z.D.; Lv, G.P. Preparation, characterization and immunoregulatory activity of derivatives of polysaccharide from *Atractylodes lancea* (Thunb.) DC. *Int. J. Biol. Macromol.* **2022**, *216*, 225–234. [CrossRef]
21. Lv, L.S.; Cheng, Y.H.; Zheng, T.S.; Li, X.M.; Zhai, R. Purification, antioxidant activity and antiglycation of polysaccharides from *Polygonum multiflorum* Thunb. *Carbohydr. Polym.* **2014**, *99*, 765–773. [CrossRef]
22. Wu, D.T.; Lam, S.C.; Cheong, K.L.; Wei, F.; Lin, P.C.; Long, Z.; Lv, X.J.; Zhao, J.; Ma, S.C.; Li, S.P. Long Simultaneous determination of molecular weights and contents of water-soluble polysaccharides and their fractions from *Lycium barbarum* collected in China. *J. Pharm. Biomed. Anal.* **2016**, *129*, 210–218. [CrossRef]
23. He, K.; Mergens, B.; Yatcilla, M.; Zheng, Q.Y. Molecular Weight Determination of Aloe Polysaccharides Using Size Exclusion Chromatography Coupled with Multi-Angle Laser Light Scattering and Refractive Index Detectors. *J. AOAC Int.* **2018**, *101*, 1729–1740. [CrossRef]
24. Chi, Y.; Li, Y.; Zhang, G.; Gao, Y.; Ye, H.; Gao, J.; Wang, P. Effect of extraction techniques on properties of polysaccharides from *Enteromorpha prolifera* and their applicability in iron chelation. *Carbohydr. Polym.* **2018**, *181*, 616–623. [CrossRef] [PubMed]
25. Tang, Y.; Zhu, Z.Y.; Liu, Y.; Sun, H.Q.; Song, Q.Y.; Zhang, Y.M. The chemical structure and anti-aging bioactivity of an acid polysaccharide obtained from rose buds. *Food Funct.* **2018**, *9*, 2300–2312. [CrossRef]
26. Wang, W.; Zou, Y.; Li, Q.; Mao, R.W.; Shao, X.J.; Jin, D.; Zheng, D.H.; Zhao, T.; Zhu, H.F.; Zhang, L.; et al. Immunomodulatory effects of a polysaccharide purified from *Lepidium meyenii* Walp. on macrophages. *Process Biochem.* **2016**, *51*, 542–553. [CrossRef]
27. Kačuráková, M.; Capek, P.; Sasinková, V.; Wellner, N.; Ebringerová, A. FT-IR study of plant cell wall model compounds: Pectic polysaccharides and hemicelluloses. *Carbohydr. Polym.* **2000**, *43*, 195–203. [CrossRef]
28. Yang, L.; Zhang, L. Chemical structural and chain conformational characterization of some bioactive polysaccharides isolated from natural sources. *Carbohydr. Polym.* **2009**, *76*, 349–361. [CrossRef]
29. Tian, W.; Dai, L.W.; Lu, S.M.; Luo, Z.F.; Qiu, Z.Y.; Li, J.J.; Li, P.; Du, B. Effect of *Bacillus* sp. DU-106 fermentation on *Dendrobium officinale* polysaccharide: Structure and immunoregulatory activities. *Int. J. Biol. Macromol.* **2019**, *135*, 1034–1042. [CrossRef]
30. Hu, X.; Pan, X.; Wang, P.; Chen, M. Isolation and characterization of an antioxidant exopolysaccharide produced by *Bacillus sp.* S-1 from Sichuan pickles. *Carbohydr. Polym.* **2019**, *204*, 9–16. [CrossRef] [PubMed]
31. Nie, C.; Zhu, P.; Ma, S.; Wang, M.; Hu, Y. Purification, characterization and immunomodulatory activity of polysaccharides from stem lettuce. *Carbohydr. Polym.* **2018**, *188*, 236–242. [CrossRef] [PubMed]
32. Cui, Y.S.; Li, Y.X.; Jiang, S.L.; Song, A.N.; Fu, Z.; Dong, C.X.; Yao, Z.; Qiao, W. Isolation, purification, and structural characterization of polysaccharidesfrom Atractylodis Macrocephalae Rhizoma and their immunostimulatory activity in RAW264.7 cells. *Int. J. Biol. Macromol.* **2020**, *163*, 270–278. [CrossRef] [PubMed]
33. Wen, L.J.; Yang, L.; Hui, L.B.; Lin, Z.C. A polysaccharide purified from Radix Adenophorae promotes cell activation and pro-inflammatory cytokine production in murine RAW264.7 macrophages. *Chin. J. Nat. Med.* **2016**, *14*, 370–376.
34. Zhou, Y.; Wang, S.C.; Feng, W.S.; Zhang, Z.L.; Li, H.W. Structural characterization and immunomodulatory activities of two polysaccharides from Rehmanniae Radix Praeparata. *Int. J. Biol. Macromol.* **2021**, *186*, 385–395. [CrossRef] [PubMed]
35. Sun, H.; Zhang, J.; Chen, F.; Chen, X.; Zhou, Z.; Wang, H. Activation of RAW264.7 macrophages by the polysaccharide from the roots of *Actinidia eriantha* and its molecular mechanisms. *Carbohydr. Polym.* **2015**, *121*, 388–402. [CrossRef]

36. Franken, L.; Schiwon, M.; Kurts, C. Macrophages: Sentinels and regulators of the immune system. *Cell. Microbiol.* **2016**, *18*, 475–487. [CrossRef]
37. Du, H.; Chen, J.; Tian, S.; Gu, H.; Li, N.; Sun, Y.; Ru, J.J.; Wang, J.R. Extraction optimization, preliminary characterization and immunological activities in vitro of polysaccharides from *Elaeagnus angustifolia* L. Pulp. *Carbohydr. Polym.* **2016**, *151*, 348–357. [CrossRef]
38. Gong, G.; Dang, T.; Deng, Y.; Han, J.; Zou, Z.; Jing, S.; Zhang, Y.; Liu, Q.; Huang, L.J.; Wang, Z.F. Physicochemical properties and biological activities of polysaccharides from *Lycium barbarum* prepared by fractional precipitation. *Int. J. Biol. Macromol.* **2018**, *109*, 611–618. [CrossRef]
39. Bogdan, C.; Ollinghoff, M.R.; Diefenbach, A. The role of nitric oxide in innate immunity. *Immunol. Rev.* **2000**, *173*, 17–26. [CrossRef]
40. Huang, L.; Shen, M.; Morris, G.A.; Xie, J. Sulfated polysaccharides: Immunomodulation and signaling mechanisms. *Trends Food Sci. Technol.* **2019**, *92*, 1–11. [CrossRef]
41. Habijanic, J.; Berovic, M.; Boh, B.; Plankl, M.; Wraber, B. Submerged cultivation of *Ganoderma lucidum* and the effects of its polysaccharides on the production of human cytokines TNF-α, IL-12, IFN-γ, IL-2, IL-4, IL-10 and IL-17. *New Biotechnol.* **2015**, *32*, 85–95. [CrossRef] [PubMed]
42. Wang, Y.F.; Tian, Y.Q.; Shao, J.J.; Shu, X.; Jia, J.X.; Ren, X.J.; Guan, Y. Macrophage immunomodulatory activity of the polysaccharide isolated from *Collybia radicata* mushroom. *Int. J. Biol. Macromol.* **2018**, *108*, 300–306. [CrossRef]
43. Li, H.; Li, J.; Shi, H.; Li, C.; Huang, W.; Zhang, M.; Luo, Y.Y.; Song, L.Y.; Yu, R.M.; Zhu, J.H. Structural characterization and immunoregulatory activity of a novel acidic polysaccharide from *Scapharca subcrenata*. *Int. J. Biol. Macromol.* **2022**, *210*, 439–454. [CrossRef] [PubMed]
44. Sun, S.; Li, K.; Xiao, L.; Lei, Z.; Zhang, Z. Characterization of polysaccharide from *Helicteres angustifolia* L. and its immunomodulatory activities on macrophages RAW264.7. *Biomed. Pharmacother.* **2019**, *109*, 262–270. [CrossRef] [PubMed]
45. Al-Banna, N.A.; Cyprian, F.; Albert, M.J. Cytokine responses in campylobacteriosis: Linking pathogenesis to immunity. *Cytokine Growth Factor Rev.* **2018**, *41*, 75–87. [CrossRef] [PubMed]
46. Liu, Q.M.; Xu, S.S.; Li, L.; Pan, T.M.; Shi, C.L.; Liu, H.; Cao, M.J.; Su, W.J.; Liu, G.M. In vitro and in vivo immunomodulatory activity of sulfated polysaccharide from *Porphyra haitanensis*. *Carbohydr. Polym.* **2017**, *165*, 189–196. [CrossRef] [PubMed]
47. Jokar, B.N.; Mehdi, T.; Guan, Y.S.; Masoud, R. Effects of extraction methods on molecular characteristics, antioxidant properties and immunomodulation of alginates from *Sargassum angustifolium*. *Int. J. Biol. Macromol.* **2017**, *101*, 703–711. [CrossRef]
48. Cao, R.A.; Lee, S.H.; You, S. Structural effects of sulfated-glycoproteins from *Stichopus japonicus* on the nitric oxide secretion ability of RAW 264.7 cells. *Prev. Nutr. Food Sci.* **2014**, *19*, 307–313. [CrossRef]
49. Rostami, Z.; Tabarsa, M.; You, S.; Rezaei, M. Relationship between molecular weights and biological properties of alginates extracted under different methods from *Colpomenia peregrina*. *Process Biochem.* **2017**, *58*, 289–297. [CrossRef]
50. Chen, Q.; Zhang, S.Z.; Ying, H.Z.; Dai, X.Y.; Ye, H.C. Chemical characterization and immunostimulatory effects of a polysaccharide from Polygoni Multiflori Radix Praeparata in cyclophosphamide-induced anemic mice. *Carbohydr. Polym.* **2012**, *88*, 1476–1482. [CrossRef]

*foods*

MDPI

*Article*

# Effects of *Quillaja* Saponin on Physicochemical Properties of Oil Bodies Recovered from Peony (*Paeonia ostii*) Seed Aqueous Extract at Different pH

**Pengkun Shen, Ruizhi Yang, Yingying Wu, Jiao Liu, Xiuzhen Ding, Wentao Wang and Luping Zhao ***

College of Food Science and Engineering, Shandong Agricultural University, Tai'an 271018, China; 17852262997@163.com (P.S.); ruizhiyang2001@163.com (R.Y.); yying1220@126.com (Y.W.); liujiao202202@163.com (J.L.); xzd@sdau.edu.cn (X.D.); wangwt@sdau.edu.cn (W.W.)

* Correspondence: zhaolp0126@126.com; Tel.: +0538-8246009

**Abstract:** Peony seeds, an important oil resource, have been attracting much attention because of α-linolenic acid. Oil bodies (OBs), naturally pre-emulsified oils, have great potential applications in the food industry. This study investigated the effects of extraction pH and *Quillaja* saponin (QS) on the physicochemical properties of peony oil body (POB) emulsions. POBs were extracted from raw peony milk at pH 4.0, 5.0, 6.0, and 7.0 (named pH 4.0-, 5.0-, 6.0-, and 7.0-POBs). All POBs contained extrinsic proteins and oleosins. The extrinsic proteins of pH 4.0- and pH 5.0-POB were 23 kDa and 38 kDa glycoproteins, the unknown proteins were 48 kDa and 60 kDa, while the 48 kDa and 38 kDa proteins were completely removed under the extraction condition of pH 6.0 and 7.0. The percentage of extrinsic proteins gradually decreased from 78.4% at pH 4.0-POB to 33.88% at pH 7.0-POB, while oleosin contents increased. The particle size and zeta potential of the POB emulsions decreased, whereas the oxidative stability, storage stability, and pI increased with the increasing extraction pH. QS (0.05~0.3%) increased the negative charges of all the POB emulsions, and 0.1% QS significantly improved the dispersion, storage, and the oxidative stability of the POB emulsions. This study provides guidance for selecting the proper conditions for the aqueous extraction of POBs and improving the stability of OB emulsions.

**Keywords:** peony oil body; extraction pH; *Quillaja* saponin; particle size; zeta potential; oxidative stability

## 1. Introduction

Oil tree peony, an important woody oil crop in China, has a long history of cultivation of around a thousand years. The oil content of peony seeds is 18.30~24.04%, and they could produce healthy and edible vegetable oil [1]. The peony seed oil contains saturated fatty acids (6.98~8.56%) and unsaturated fatty acids (91.44~93.02%), among which α-linolenic acid accounts for more than 40% [2]. It was found that vegetable oil existed in the form of natural lipid-storing organelles called oil bodies in rapeseed, beans, and nuts [3]. Oil bodies (OBs) are mainly composed of triglycerides (TAGs, 94.2~98.2%), oleosins (0.3~3.0%), and phospholipids (PL, 0.6~2.0%) [4]. OBs have a TAG matrix core coated by one monolayer of phospholipids embedded with structural proteins. Oleosins play crucial roles in keeping the structural integrity of OBs and resisting some external environmental stresses [5]. Based on the above structural characteristics of OBs and peony seed oil, peony oil bodies (POBs) have a potential broad application in both the food and cosmetics industries.

OBs with different components (such as proteins, neutral lipids, and phospholipids) were extracted by varying the extraction pH. Sun et al. (2023) [6] extracted peanut OBs at pH 4.0, 7.5, and 11.0, with the moisture and fat contents gradually increasing with the increasing pH, while the protein content decreased from 1.47% to 1.01%. Moreover, different extraction pH also affected the properties of the OB emulsions. Chen et al. (2014) [7] found that the neutral lipid percentages of soybean OBs increased from 85.88% to 91.89%

when increasing the extraction pH from 6.8 to 11.0, while the percentage of proteins and polar lipids decreased. It was found that peanut OB emulsions extracted at pH 7.5 and pH 4.0 had smaller particle sizes, higher interfacial viscoelasticity, and a better emulsification performance than peanut OB emulsions extracted at pH 11.0 [6]. At present, Zhao et al. (2020) [8] extracted POBs at pH 7.0, with the stability of the POB decreasing by adding 50~100 mM NaCl, but the POBs were still good at 30~90 °C. However, the influence of the different extraction pH on the composition and properties of POB emulsions has not been systematically studied. In addition, it is important that OB emulsions have a good physicochemical stability for their applications in food. Polysaccharides and polyphenols were usually used to improve the physicochemical properties of OB emulsions in recent years. Interestingly, saponin, a natural small molecule surfactant, has excellent properties in reducing the interfacial tension of oil and water due to its amphiphilicity and improved emulsion stability. It was found that the particle size of the emulsion prepared with soy lecithin and medium chain triglycerides decreased from 0.36 μm to 0.15 μm by increasing the *Quillaja* saponin (QS) content from 0.5% to 2.5% [9]. Zhu et al. (2019) [10] prepared stable nano-emulsions with 1% tea saponins and found that the tea saponins could reduce the interfacial tension from 24.42 mN/m to 4.8 mN/m, improving storage stability. Therefore, saponins are effective in improving the common emulsion stability. However, only Liu et al. (2022) [11] found that 0.1% and 1.0% camellia saponins effectively delayed the lipid oxidation of the camellia OB emulsion extracted at pH 7.0. More research needs to be carried out in order to realize the effects of saponins on OB emulsions.

In this study, we extracted POBs from peony seeds at pH 4.0, 5.0, 6.0, and 7.0, and their compositions and isoelectric points were determined. The effects of QS (0.05~0.3%, *w*/*w*) on the particle size, zeta potential, microstructure, storage stability, and oxidative stability on the different POB emulsions were researched. This will provide theoretical guidance for improving OB emulsions and POB applications in food.

## 2. Materials and Methods

### 2.1. Materials

The peony seeds (*Paeonia ostia*, 10 kg) were obtained from Shandong Seed Industry Group CO., Ltd. (Jinan, China). *Quillaja* saponin was purchased from Shanxi Haosen Biological Technology Co., Ltd. (Xi'an, China). Protein Markers (14.4~97.4 kDa) were bought from Beijing Solarbio Science and Technology Co., Ltd. (Beijing, China) Mercaptoethanol, Bromophenol blue, Coomasil bright blue, glacial acetic acid, isoctane, isopropyl alcohol, methanol, n-butanol, thiobarbituric acid, sodium azide, trichloroacetic acid, and trichloromethane were purchased from Sinopharm Chemical Reagent Co., Ltd. (Shanghai, China). All reagents were of analytical grade.

### 2.2. POBs Extracted at Different pH

The extraction method of the POBs was performed with reference to the previous method by Nikiforidis et al. (2009) [12]. The peony seed kernels were cleaned with deionized water, soaked with deionized water in a beaker, which was then placed in a refrigerator at 4 °C for 16 h. The soaked kernels were mixed with deionized water (kernels/deionized water = 1/7, *w*/*w*), and ground for 2 min by a blender (18,000 r/min; Blst4090-073, Qster, Qingdao, China). The slurry was filtered using four layers of gauze, and the raw peony milk was collected, which was then adjusted to the required pH 4.0, 5.0, 6.0, 7.0, and 11.0. The mixtures were centrifuged at 13,433× *g* (4 °C, 30 min). The precipitate extracted at pH 4.0 was collected from the bottom of centrifugal tube; however, the floats extracted at pH 5.0, 6.0, 7.0, and 11.0 were collected from the upper layer. The collections were dispersed into DI water again, and the pH of the mixtures were adjusted to 4.0, 5.0, 6.0, 7.0, and 11.0. The mixtures were centrifuged in the same conditions as above. The precipitate extracted at pH 4.0 was collected and named pH 4.0-POB, the floats extracted at pH 5.0, 6.0, 7.0, and 11.0 were collected and named pH 5.0-, 6.0-, 7.0, 11.0-POB, respectively.

### 2.3. Measurement of POB Composition

The moisture content was measured using the oven-drying method, and the temperature was controlled at 105 °C. The measurement of the protein content was based on the method of Pan et al. (2022) [13]. Here, the coefficient of protein converted by nitrogen was 6.25. The lipid content was determined using Soxhlet extraction [14].

### 2.4. Tricine-SDS-PAGE

Samples were mixed with the sample loading buffer (containing 0.1% ($v/v$) bromophenol blue and 2% ($v/v$) mercaptoethanol). The protein concentrations of all samples were adjusted to 1.5 mg/mL via mixing with sample loading buffer. The mixed solutions were heated using a boiling water bath for 5 min, and they were centrifuged at 10,000× $g$ for 5 min. Then, 10 μL of each sample was injected into the sample well. Electrophoresis was performed using a Mini-Protean instrument (DYY-8C, Beijing Liuyi Biotechnology Co., Ltd., Bejing, China) at a constant voltage of 30 mV until the mixed solution entered into the stacked gel; then, the voltage was changed to 100 mV until complete. The gel was fixed with 50% ($v/v$) methanol and 10% ($v/v$) acetic acid, and they were stained with 0.025% ($w/w$) Coomassie blue and de-colorized with 10% ($v/v$) acetic acid. The Image Lab 3.0 Software (Bio-Rad, Hercules, CA, USA) was used to take photos and analyze the intensities of the different protein bands.

### 2.5. Emulsion Preparation

The pH 4.0-, 5.0-, 6.0-, and 7.0-POBs were dispersed into deionized water to obtain the POB emulsions (10%, $w/w$), and their pH were adjusted to 7.0, respectively, using 0.1~2 M NaOH solutions. Each POB emulsion was, respectively, added QS (0.05%, 0.1%, 0.2%, 0.3%; $w/w$), and all samples were mixed well. To prevent microbial growth, 0.02% ($w/v$) sodium azide was added into all POB emulsions and the follow-up analyses were conducted immediately.

### 2.6. Zeta Potential

The pHs of the different POB emulsions (10%, $w/w$) were, respectively, adjusted to 3.0, 4.0, 5.0, 6.0, 7.0 with 0.01 M NaOH and 0.01 M HCl solutions, and they were diluted 400 times with same pH buffers (phosphate buffer and citric acid buffer; 20 mM). Zeta potentials were determined using a Laser Nanometer Particle Size Analyzer (Malvern Instruments Ltd., London, UK) at 25 °C.

### 2.7. Particle Size

All POB emulsions were diluted 1000 times using a phosphate-buffered solution (20 mM, pH 7.0), respectively. The particle sizes were determined using a Laser Nanometer Particle Size Analyzer (Malvern Instruments Ltd., London, UK) at 25 °C.

### 2.8. Microstructures

Microstructures were observed and images were recorded at 25 °C using an optical microscope (MODEL BX-51, TKO Optical Instruments Co., Ltd., Tokyo, Japan). Samples (10 μL) were deposited on a microscope slide and covered with a cover slip. Samples were observed with a 10× eyepiece and a 100× objective lens.

### 2.9. Determination of Oxidative Stability

Fresh 10% POB emulsions (pH 7.0), were stored at 37 °C with restricted lighting in closed glass vials for different time periods up to 14 days according to the method of Zhao et al. (2016) [15].

#### 2.9.1. Lipid Hydroperoxides

The peroxide value (POV) was determined using the modified methods described by Salminen et al. (2014) [16]. Each POB emulsion (0.3 mL) was mixed with 1.5 mL isooctane–isopropanol solution ($v/v = 2/1$). Each mixture was shaken twice for 20 s, and it was centrifuged at 900× $g$ for 10 min. The supernatant (0.2 mL) was mixed with KSCN solution (20 μL, 3.94 M) and Fe2+ solution (20 μL, 0.072 M), and the volume was made to 10 mL with methanol–n-butanol solution ($v/v$ = 2/1), which was kept in the dark for 20 min. The absorbance was measured at 510 nm using a UV-5100 spectrophotometer (Shanghai Yuan Analysis Instrument Co., Ltd., Shanghai, China), and the methanol–n-butanol solution was used as control. All peroxide values were calculated according to the standard curve using the following formula:

$$\mathrm{POV(mmol/kg)} = \frac{\mathrm{M} \times \mathrm{A}_{510}}{55.84 \times 2 \times \mathrm{m}_0}$$

$A_{510}$ represents the absorbance value of 510 nm; M represents the slope of the $Fe^{3+}$ standard curve; $m_0$ represents the colorimetric equivalent sample mass, g.

2.9.2. Thiobarbituric Acid Reactive Substances (TBARS)

The secondary products of lipid oxidation were evaluated by TBARS according to the method by Jiang et al. (2022) [17], albeit with modifications. Thiobarbituric acid solution (TBA) was prepared by dissolving thiobarbituric acid (93.75 mg) and trichloroacetic acid (3.75 g) into 25 mL HCl (0.25 M). An amount of 0.5 mL of each emulsion was mixed with TBA reagent (1 mL) in a test tube, heated in a boiling water bath for 15 min, and cooled to room temperature. The mixture was added into 0.3 mL of trichloromethane, and it was centrifuged at 3000× $g$ for 15 min. The supernatant was collected and the absorbance was measured at 532 nm. The TBARS content was calculated using a standard curve prepared from 1, 1, 3, 3-tetroxypropane (0~30 μM).

### 2.10. Calculation and Statistical Analysis

All POB emulsions were prepared three times on different days. Values of the different parameters were presented as the mean ± standard deviations. Mean and standard deviations were obtained by Origin Pro 2016b (OriginLab Co., Northampton, MA, USA). Statistical analysis was performed using SPSS statistics version with 25 ANOVA (IBM Inc., Armonk, NY, USA). The means were compared by a least significant difference (LSD) test at the 5% level ($p$ = 0.05).

## 3. Results and Discussion

### 3.1. Component Analysis of POBs

Table 1 showed that the contents of moisture, protein, and lipid of pH 4.0-, 5.0-, 6.0-, and 7.0-POBs. The pH 4.0-POB contained the highest moisture content (52.39%), followed by pH 5.0-POB (46.14%) and pH 6.0-POB (28.75%), and pH 7.0-POB had the lowest moisture content (20.98%). The protein contents of the different POBs decreased from 33.40% to 4.27%, while the lipid contents increased from 57.60% to 86.31% with increasing extraction pH from 4.0 to 7.0. This pattern was consistent with peanut OBs and soybean OBs extracted at pH 6.8~11.0 [15,18]. Considering the ability of proteins to bind water, it was believed that the water contents of the different OBs increased due to the increasing of protein contents [19].

**Table 1.** Contents of moisture, protein, and lipid of pH 4.0-, 5.0-, 6.0-, and 7.0-POB. Different small letters represent significant differences ($p$ < 0.05) in the same component.

| Samples | Compositions | | |
|---|---|---|---|
| | Moisture (%) | Lipid (%, Dry Basis) | Protein (%, Dry Basis) |
| pH 4.0-POB | 52.39 ± 0.69 [a] | 57.60 ± 3.45 [c] | 33.40 ± 0.37 [a] |
| pH 5.0-POB | 46.14 ± 1.30 [b] | 76.76 ± 0.93 [b] | 13.57 ± 1.49 [b] |
| pH 6.0-POB | 28.75 ± 1.07 [c] | 83.52 ± 2.18 [ab] | 4.82 ± 0.13 [c] |
| pH 7.0-POB | 20.98 ± 1.25 [d] | 86.31 ± 4.16 [a] | 4.27 ± 0.54 [c] |

Figure 1A showed the protein composition of POBs, and the protein types gradually decreased with increasing extraction pH. In order to determine the oleosins, we extracted the high purity pH 11.0-POB. Ishii et al. (2021) [20] reported that pH 11.0 was a simple method for obtaining high purity OBs. It was found that pH 11.0-POB contained three proteins (15 kDa,

17 kDa and 23 kDa). According to Zhao et al. (2020) [8] and Meng et al. (2020) [21], there were oleosin 15.06 kDa and oleosin 17.5 kDa in peony seeds. Therefore, 15 kDa and 17 kDa in electrophoretic gel were probable oleosins. In addition, Gao et al. (2018) [22] found that peony seed storage proteins contained 60 kDa, 48 kDa, 38 kDa and 23 kDa, respectively, and 38 kDa and 23 kDa were glycoproteins with isoelectric points (pIs) of 3.6 and 9.0. Zhao et al. (2016) [15] reported that storage proteins were released from protein storage vacuoles during grounding seeds, and some of them were adsorbed to the surface of soybean OBs, which were called extrinsic proteins of soybean OBs. Therefore, all POBs contained not only the oleosins (15 and 17 kDa) but also the extrinsic proteins. Both pH 4.0- and pH 5.0-POB contained four extrinsic proteins (23 kDa and 38 kDa glycoproteins, unknown proteins 48 kDa and 60 kDa), while 38 kDa glycoprotein and unknown 48 kDa were completely removed when the extraction pH was more than 6.0. This was probably because the extraction pH 6.0 and 7.0 were much higher than the pI 3.6 of 38 kDa glycoprotein, and the electrostatic repulsion between POBs and 38 kDa glycoprotein increased, which resulted in 38 kDa glycoprotein releasing from POBs. Interestingly, 5.60% 23 kDa glycoprotein was still bounded to pH 11.0-POB. It was attributed to the weaker electrostatic repulsion between POBs and glycoprotein 23 kDa (pI 9.0), and it is difficult to completely remove 23 kDa glycoprotein. Similarly, 11S globulin was removed from sesame OBs at extraction pH 8.0, whereas 2S albumin was completely removed at the extraction pH 11.0 [23]. It was likely that the interaction forces between the different extrinsic proteins and OBs were different. Zhao et al. (2013) [24] reported that the extrinsic Gly m Bd 30K and P34 probable thiol protease were strongly bound to soybean OBs through the disulfide bond with 24 kDa oleosin, and Gly m Bd 30K and P34 probable thiol protease were difficult to remove from soybean OBs. The analysis of protein band strength showed that the extrinsic proteins gradually decreased with increasing extraction pH from 4.0 to 7.0 (Figure 1B). The percentage of the extrinsic proteins gradually reduced from 78.4% of pH 4.0-POB to 33.88% of pH 7.0-POB, while the oleosin contents increased.

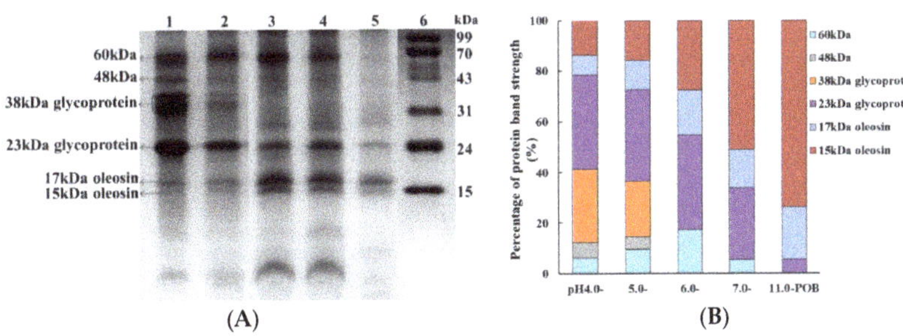


**Figure 1.** (**A**) Tricine-SDS-PAGE of POBs extracted at different pH. Lanes 1~5, pH 4.0-, 5.0-, 6.0-, 7.0- and 11.0-POB; Lane 6, protein marker. (**B**) Percentages of 60 kDa, 48 kDa, glycoprotein 38 kDa, glycoprotein 23 kDa, and oleosins in total proteins of pH 4.0-, 5.0-, 6.0-, 7.0- and 11.0-POB.

### *3.2. Zeta Potential*

The surface charges of the POBs were affected by environmental pH. As the environmental pH increased, the surface charges of the POBs decreased gradually from positive at pH 3.0~pH 4.0 to negative at pH 5.0~pH 7.0 (Figure 2A). The droplet charges were 0 at certain pH, which were the pI. The pIs of the pH 4.0-, 5.0-, 6.0-, and 7.0-POB were approximately 4.2, 4.3, 4.5, and 4.6, respectively. The pIs of POBs showed an increasing tendency with increasing extraction pH, which is consistent with the results reported by Zhao et al. (2016) [23]. In addition, pIs of pH 4.0- and pH 5.0-POB were close to their extraction pH, which resulted in more extrinsic proteins bounding to pH 4.0- and pH 5.0-POBs with the weaker electrical repulsions. This explained why the extrinsic protein of POBs gradually decreased with increasing extraction pH from 4.0 to 7.0 (Figure 1).

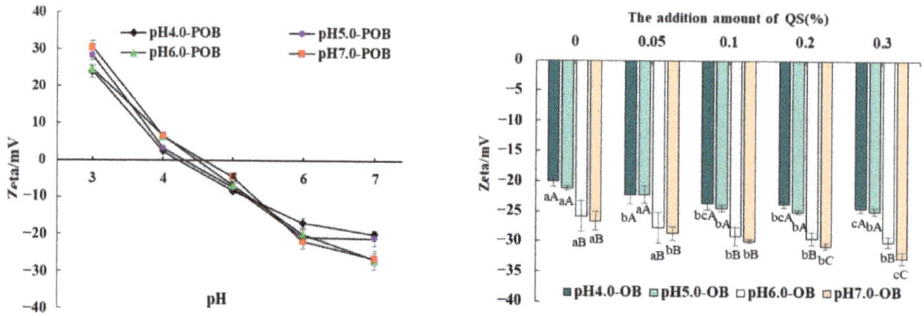


(A) (B)

**Figure 2.** (**A**) Zeta potential of pH 4.0-, 5.0-, 6.0-, and 7.0-POBs at the different environment pH. (**B**) Effect of QS (0~0.3%, $w/w$) on the zeta potential of pH 4.0-, 5.0-, 6.0-, and 7.0-POB emulsions. Different capital or small letters represent significant differences ($p < 0.05$).

Figure 2B showed the effect of QS on the zeta potential of the different POB emulsions (pH 7.0). The zeta potential of the pH 4.0, 5.0-, 6.0, and 7.0-POB emulsions without QS were −20.03 mV, −21.15 mV, −25.87 mV, −26.63 mV, respectively. The addition of QS, especially 0.1~0.3%, decreased the zeta potential of each POB emulsion (pH 7.0). It was attributed to the possibility that QS with negative charge might adsorb to the POB surface. Yang et al. (2013) [25] reported that QS had carboxylic acid groups (pKa values around pH 3.5), and the groups would be fully charged (-COO-) at high pH values (pH > pKa). In addition, Liu et al. (2022) [11] reported that camellia saponins resulted in decreasing the zeta potential of camellia OB emulsions because camellia saponins and camellia OB interfacial proteins formed the complex. Under the same contents of QS, the zeta potential of the POB emulsions generally showed a decreasing trend with increasing extraction pH. The pH 4.0- and pH 5.0-POB emulsions had less charge than those of pH 6.0- and pH 7.0-POB at the same QS addition. It might be related that the initial charges of pH 4.0- and pH 5.0-POB emulsions were less (Figure 2B). Therefore, the extraction pH decreased the pIs of POBs, and the QS increased the negative charges of POB emulsions.

### 3.3. Particle Size

Without QS, the initial particle sizes of pH 4.0-, 5.0-, 6.0-, and 7.0-POB emulsions were 1280.50 nm, 1178.50 nm, 718.03 nm and 709.53 nm, respectively (Figure 3A). It was due to the different amounts of extrinsic proteins that were bound to POBs. It was indicated that the initial particle sizes decreased with increasing extraction pH. However, the addition of QS could decrease the particle sizes of all POB emulsions. Under the same extraction pH, the particle size decreased firstly, and then increased with increasing QS concentration from 0.05% to 0.3%. On the one hand, more QS molecules with negative charges could absorb to POB interfacial with increasing QS from 0.05% to 0.1%, which prevented droplet aggregation by increasing the electrical repulsion, which decreased the particle sizes. On the other hand, the excessive QS (more than 0.1%) might combine with POB interfacial proteins to form the complex that dissolved in the bulk phase. Li et al. (2023) [26] reported that there was an appropriate concentration of saponins combined with proteins in a certain ratio that would work best, while excessive saponins would occupy the surface layer, which affected the properties of emulsions. At the QS concentration of 0.1%, the particle sizes of pH 4.0-, 5.0-, 6.0-, and 7.0-POB emulsions were smallest, and were 774.73 nm, 620.95 nm, 341.00 nm and 398.40 nm, respectively. The particle size distributions of pH 4.0-, 5.0-, 6.0-, and 7.0-POB emulsions with 0.1% QS are shown in Figure 3B. The pH 4.0-, and 5.0-POB emulsions with 0.1% QS presented a bimodal distribution with a wide particle size distribution, while other emulsions showed a unimodal distribution. The main peak position moved toward the direction of small droplets. Therefore, the particle sizes of POB emulsions decreased with increasing extraction pH, and 0.1% QS had better effects on decreasing the particle sizes.

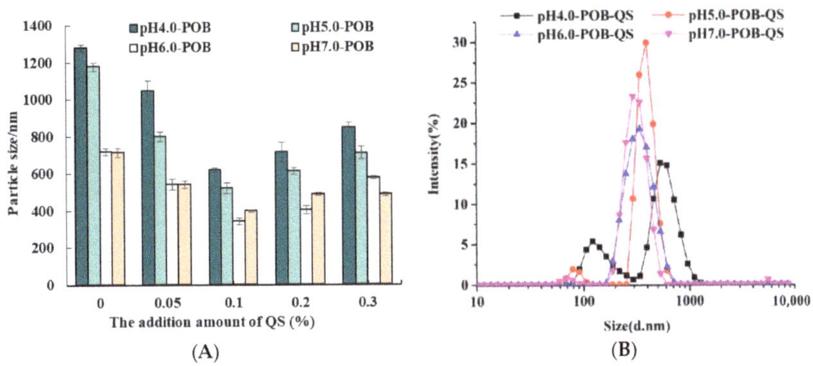


**Figure 3.** (**A**) Effect of QS (0~0.3%, *w/w*) on the particle size of pH 4.0-, 5.0-, 6.0-, and 7.0-POB emulsions. (**B**) The size distributions of pH 4.0-, 5.0-, 6.0-, and 7.0-POB emulsions with 0.1% (*w/w*) QS.

### 3.4. Microstructure

The microscope images showed that some larger droplets and numerous aggregates (marked with circles in Figure 4) were obviously observed in initial pH 4.0- and 5.0-POB emulsions on the first day, while initial pH 6.0- and 7.0-POB emulsions had small and few aggregates. The aggregation was another reason for increasing the particle sizes of pH 4.0- and 5.0-POBs emulsions without QS (Figure 3A). When QS was added, the aggregates decreased and the droplets in POB emulsions were dispersed as individuals, especially pH 4.0- and 5.0-POB emulsions. This result was consistent with Figure 3, and it was due to the increase in negative charges and electrical repulsion at the presence of QS (Figure 2B). It was revealed that QS could improve the dispersion of POB emulsions. It was shown that the droplets in all emulsions with 0.1% QS was smallest and dispersed as individuals. In addition, some individual large droplets (marked with red arrows) were still observed in pH 4.0- and 5.0-POB emulsions with 0.05% and 0.3% QS. It was possible that the droplets were not completely covered with less QS, while excessive QS induced interfacial proteins released from POBs, and the QS-induced coalescence of POBs occurred. These results revealed that POB emulsions extracted at neutral pH had a better dispersion stability compared to the acid-extraction pH, and 0.1% QS had a better effect on improving dispersion stability of the POB emulsions.

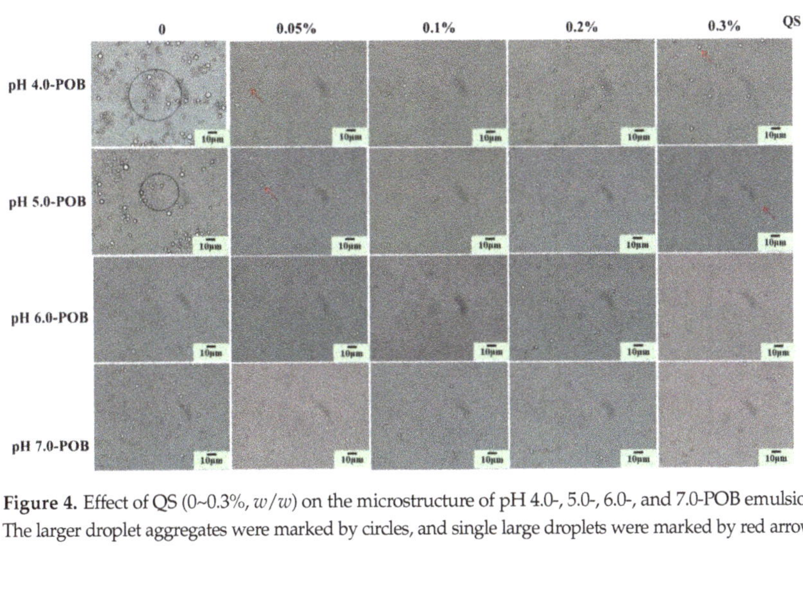


**Figure 4.** Effect of QS (0~0.3%, *w/w*) on the microstructure of pH 4.0-, 5.0-, 6.0-, and 7.0-POB emulsions. The larger droplet aggregates were marked by circles, and single large droplets were marked by red arrows.

### 3.5. Storage Stability

The storage stability of emulsions is very important in food and beverage applications. The storage stability of the different POB emulsions was analyzed using macroscopic images and particle sizes during storage for 14 days. Figure 5 showed that the clear serum layer was observed in only pH 4.0-POB emulsions without QS during 14 days of storage. The creams (marked with rectangles) were observed on the top of pH 4.0-POB emulsions after 7 days of storage, and the presence of QS considerably decreased the thickness of creams. It was attributed that the droplets in pH 4.0-POB emulsions formed more and larger aggregates during storage for 7~14 days, and the addition of QS inhibited droplet aggregation. Similarly, the unclear creams (marked with rectangles) also appeared on the top of all pH 5.0-POB emulsions after 14 days of storage. However, no creams formed on the top of all pH 6.0-, and 7.0-POB emulsions during storage, which was suggested in the good dispersion. It was relative to more charges and smaller particle sizes of initial pH 6.0-, and 7.0-POB emulsions (Figures 2B and 4). It can be seen that the storage stability of POB emulsions was improved with increasing extraction pH, which might be relative to the extrinsic proteins of POBs. Ishii et al. (2017) [27] reported that more extrinsic proteins of soybean OB resulted in a greater degree of particle aggregation.

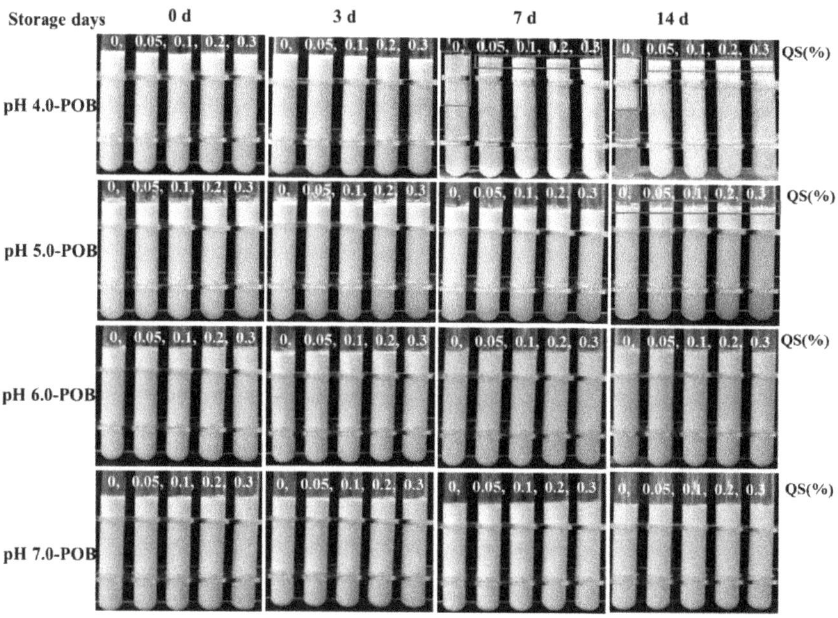


**Figure 5.** Effect of QS (0~0.3%, *w*/*w*) on the visual appearance of pH 4.0-, 5.0-, 6.0-, and 7.0-POB emulsions during storage for 0, 3, 7, 14 days. Creams were marked by rectangles.

Figure 6 shows that the particle sizes of all POB emulsions increased with prolonged storage days, and QS obviously decreased their particle sizes. Without QS, the particle sizes of pH 4.0-, 5.0-, 6.0-, and 7.0-POB emulsions increased to 3185.67, 2833.67, 1708.00 and 1699.50 nm after storage for 14 days, respectively. It was revealed that the higher pH was used to extract the POB, while the smaller aggregates existed in POB emulsions, which is consistent with the macroscopic images in Figure 5. In particular, the particle sizes of pH 6.0- and 7.0-POB emulsions without QS had no significant difference ($p < 0.05$) at storage for 7 and 14 days, which is attributed to the similar content and composition of POB proteins. During storage, the particle sizes of all POB emulsions obviously decreased with increasing the QS concentration from 0.05% to 0.1%, then they increased slightly with the QS increase from 0.2% to 0.3%. Under the condition of 0.1% QS, pH 6.0- and 7.0-POB emulsions had the smallest particle sizes and better storage stability than other emulsions.

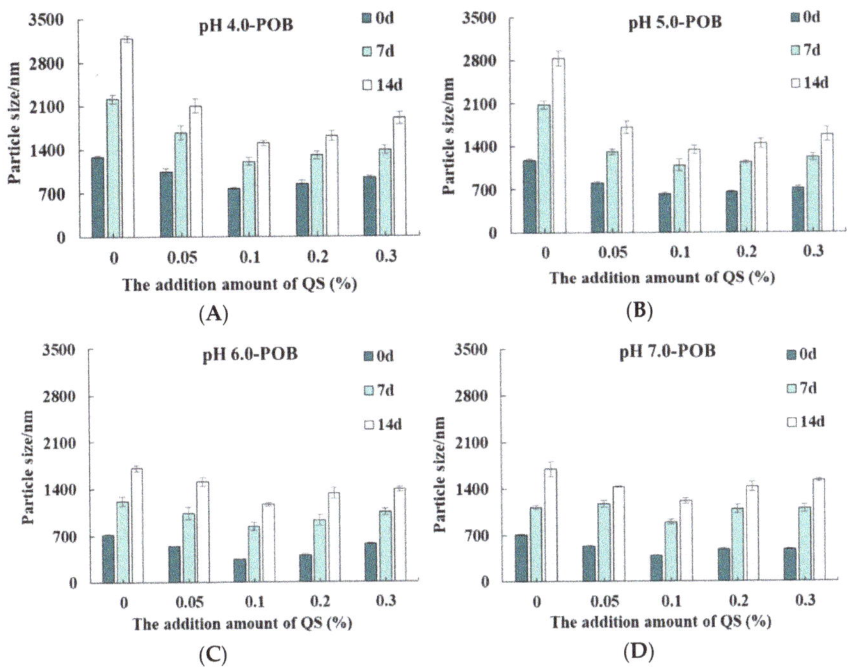


**Figure 6.** Effect of QS content (0~0.3%, $w/w$) on the particle size of pH 4.0-(**A**), 5.0-(**B**), 6.0-(**C**), and 7.0-POB emulsions (**D**) during storage.

### 3.6. Oxidative Stability

Oxidative stability refers to the ability of the emulsions to resist oxidative acidification and maintain the stability of the system [28]. In order to evaluate the oxidation stability of POB, the POVs (primary oxidation products) and TBARS (secondary oxidation products) were regularly monitored. Figure 7A showed that POV contents (the primary lipid oxidation product) increased first, and then decreased with prolonged storage days in all POB emulsions. It was attributed to the conversion of some primary oxidation products to secondary oxidation products [29]. In the absence of QS, the POV increased in 0~4 days, and decreased in 4~14 days, and the highest POV of pH 4.0-, 5.0-, 6.0-, and 7.0-POB emulsions were, respectively, 4.7, 4.28, 3.8, 3.93 mmol/kg at the fourth storage day. When 0.1% ($w/w$) QS was added into the POB emulsions, all POVs decreased, which suggested that QS could inhibit the oxidation of lipids. This might be attributed to the following aspects: first, the addition of saponins increased the amount of phenolic compounds, leading to more hydrogen donors to reduce free radicals [30]; and second, droplet charges increased by QS accumulating at the POB interface (Figure 2), which might improve the interface physical barrier and inhibit the contact between water-soluble components and lipids. As shown in Figure 7B, TBARS content of all POB emulsions gradually increased prolonged storage days at similar rates. Under the same extraction pH of POB emulsions, the TBARS of POB emulsions without QS were higher, and it was revealed QS also decreased the produce rate of secondary oxidation products. In addition, the POV and TBARS in pH 6.0- and 7.0-POB emulsions were lower than those of pH 4.0- and 5.0-POB emulsions at the same concentration of QS, which might be because pH 6.0- and 7.0-POB emulsions had more charges (Figure 2) and less extrinsic proteins (Figure 1B). Zhao et al. (2016) [15] reported that extrinsic proteins of soybean OBs played a crucial role in accelerating the oxidation of lipids, and oxidative stability increased with decreasing extrinsic proteins. Therefore, QS (0.1%, $w/w$) improved the oxidation stability of POB emulsions, and POB emulsions extracted at higher pH had better oxidation stability.

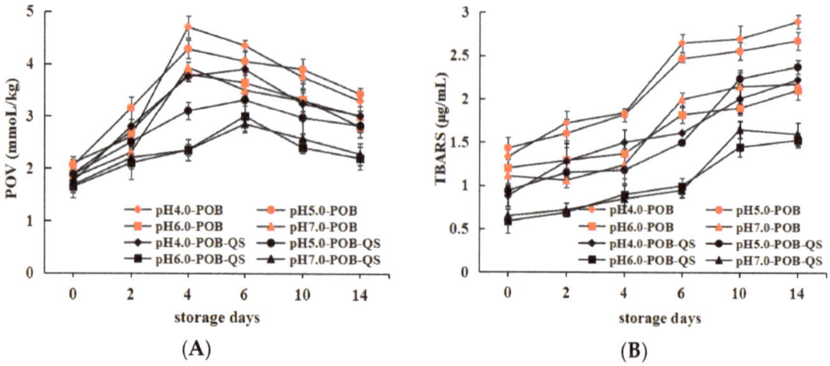


**Figure 7.** POV (**A**) and TBARS (**B**) of pH 4.0-, 5.0-, 6.0- and 7.0-POB emulsions without QS and with 0.1% ($w/w$) QS as a function of storage days.

## 4. Conclusions

In this study, we prepared emulsions with POBs extracted at pH 4.0, 5.0, 6.0, 7.0, and discussed the effects of QS on physicochemical properties of POB emulsions. It was found that moisture and protein contents decreased, while lipid content increased with increasing extraction pH. All POBs had 15 kDa oleosin and 17 kDa oleosin, while the different POBs had different extrinsic proteins. At the lower extraction pH (4.0 and 5.0), extrinsic proteins (23 kDa and 38 kDa glycoproteins, unknown proteins 48 kDa and 60 kDa) were bounded to POBs, while 48 kDa and 38 kDa were completely removed at extraction pH 6.0 and 7.0. The particle sizes and zeta potential of POB emulsions decreased, whereas pI, storage stability and oxidative stability increased in the order of pH 4.0-, 5.0-, 6.0- and 7.0-POB. QS (0.05~0.3%, $w/w$) increased the negative charges of POB emulsions, decreased the particle sizes, and inhibited the droplet aggregation and oxidation of lipids. 0.1% QS could significantly improve the dispersion, storage and oxidative stability of POB emulsions. Therefore, extraction pH and QS had important effects on the physicochemical properties of POB emulsions. It was considered that the pH 6.0- and 7.0-POB emulsions with 0.1% QS could be used for plant beverages. This study will provide a strategy for improving the stability of OB emulsions and the application of POBs in the food industry.

**Author Contributions:** Conceptualization, L.Z.; Methodology, P.S. and R.Y.; Software, R.Y.; Validation, Y.W. and J.L.; Formal analysis, Y.W., J.L. and L.Z.; Investigation, P.S. and R.Y.; Resources, X.D. and W.W.; Data curation, L.Z.; Writing—original draft, P.S.; Writing—review and editing, L.Z.; Visualization, X.D. and W.W.; Supervision, X.D. and W.W.; Project administration, L.Z.; Funding acquisition, L.Z. All authors have read and agreed to the published version of the manuscript.

**Funding:** This research was funded by the Natural Science Foundation of Shandong Province, PR China (No. ZR2022QC189).

**Data Availability Statement:** The data presented in this study are available on request from the corresponding author.

**Conflicts of Interest:** The authors declare no conflict of interest.

## References


1. Deng, R.; Gao, J.; Yi, J.; Liu, P. Peony seeds oil by-products: Chemistry and bioactivity. *Ind. Crops Prod.* **2022**, *187*, 115333. [CrossRef]
2. Deng, R.; Gao, J.; Yi, J.; Liu, P. Could peony seeds oil become a high-quality edible vegetable oil? the nutritional and phytochemistry profiles, extraction, health benefits, safety and value-added-products. *Food Res. Int.* **2022**, *156*, 111200. [CrossRef]
3. Zaaboul, F.; Zhao, Q.; Xu, Y.; Liu, Y. Soybean oil bodies: A review on composition, properties, food applications, and future research aspects. *Food Hydrocoll.* **2022**, *124*, 107296.
4. Tzen, J.T.C.; Cao, Y.; Laurent, P.; Ratnayake, C.; Huang, A.H.C. Lipids, proteins, and structure of seed oil bodies from diverse species. *Plant Physiol.* **1993**, *101*, 267–276. [CrossRef] [PubMed]
5. Shimada, T.L.; Hara-nishimura, I. Oil-body-membrane proteins and their physiological functions in plants. *Biol. Pharm. Bull* **2010**, *33*, 360–363. [CrossRef]
6. Sun, F.; Wang, Q.; Gao, C.; Xiao, H.; Yang, N. Effect of extraction pH and post-extraction heat treatment on the composition and interfacial properties of peanut oil bodies. *Colloids Surf. A Physicochem. Eng. Asp.* **2023**, *656*, 130351. [CrossRef]
7. Chen, Y.; Cao, Y.; Zhao, L.; Kong, X.; Hua, Y. Macronutrients and micronutrients of soybean oil bodies extracted at different pH. *J. Food Sci.* **2014**, *79*, 1285–1291. [CrossRef]
8. Zhao, D.; Li, T.; Li, Z.; Sun, J.; Tao, J. Characteristics of *Paeonia ostii* seed oil body and OLE 17.5 determining oil body morphology. *Food Chem.* **2020**, *319*, 126548. [CrossRef]
9. Chung, C.; Sher, A.; Rousset, P.; Decker, E.A.; McClements, D.J. Formulation of food emulsions using natural emulsifiers: Utilization of quillaja saponin and soy lecithin to fabricate liquid coffee whiteners. *J. Food Eng.* **2017**, *209*, 1–11.
10. Zhu, Z.; Wen, Y.; Yi, J.; Cao, Y.; Liu, F.; McClements, D.J. Comparison of natural and synthetic surfactants at forming and stabilizing nanoemulsions: Tea saponin, Quillaja saponin, and Tween 80. *J. Colloid Interface Sci.* **2019**, *536*, 80–87. [CrossRef] [PubMed]
11. Liu, J.; Hu, L.; Chen, Y.; Xiao, Y.; Zheng, M.; Yu, Z.; Liu, Y.; Zhou, Y. Effects and mechanism of camellia saponin on the physicochemical and oxidative stability of camellia oil body-based emulsions. *LWT-Food Sci. Technol.* **2022**, *165*, 113773. [CrossRef]
12. Nikiforidis, C.V.; Kiosseoglou, V. Aqueous extraction of oil bodies from maize germ (*Zea mays*) and characterization of the resulting natural oil-in-water emulsion. *J. Agric. Food Chem.* **2009**, *57*, 5591–5596. [CrossRef] [PubMed]
13. Pan, Y.; Jin, W.; Huang, Q. Structure, assembly and application of novel peanut oil body protein extracts nanoparticles. *Food Chem.* **2022**, *367*, 130678. [CrossRef]
14. Zhang, S.; Chen, H.; Geng, F.; Peng, D.; Xie, B.; Sun, Z.; Chen, Y.; Deng, Q. Natural oil bodies from typical oilseeds: Structural characterization and their potentials as natural delivery system for curcumin. *Food Hydrocoll.* **2022**, *128*, 107521. [CrossRef]
15. Zhao, L.; Chen, Y.; Yan, Z.; Kong, X.; Hua, Y. Physicochemical and rheological properties and oxidative stability of oil bodies recovered from soybean aqueous extract at different pHs. *Food Hydrocoll.* **2016**, *61*, 685–694. [CrossRef]
16. Salminen, H.; Aulbach, S.; Leuenberger, B.H.; Tedeschi, C.; Weiss, J. Influence of surfactant composition on physical and oxidative stability of Quillaja saponin-stabilized lipid particles with encapsulated ω-3 fish oil. *Colloids Surf. B Biointerfaces* **2014**, *122*, 46–55. [CrossRef]
17. Jiang, T.; Charcosset, C. Encapsulation of curcumin within oil-in-water emulsions prepared by premix membrane emulsification: Impact of droplet size and carrier oil on the chemical stability of curcumin. *Food Res. Int.* **2022**, *157*, 111475.
18. Zaaboul, F.; Raza, H.; Chen, C.; Liu, Y. Characterization of Peanut Oil Bodies Integral Proteins, Lipids, and Their Associated Phytochemicals. *J. Food Sci.* **2017**, *83*, 93–100. [CrossRef]
19. Xu, D.; Gao, Q.; Ma, N.; Hao, J.; Yuan, Y.; Zhang, M.; Cao, Y.; Ho, C.-T. Structures and physicochemical characterization of enzyme extracted oil bodies from rice bran. *LWT-Food Sci. Technol.* **2021**, *135*, 109982.
20. Ishii, T.; Matsumiya, K.; Matsumura, Y. Combinational effects of acid and salt addition on colloidal, interfacial, and emulsifying properties of purified soybean oil bodies. *Food Hydrocoll.* **2021**, *111*, 106213. [CrossRef]
21. Meng, J.; Tang, Y.; Zhang, X.; Zhao, D.; Tao, J. Properties of oil body development and characteristics of *OLE* gene in herbaceous peony (*Paeoinia lactiflora* Pall.) seeds. *J. Hortic. Sci. Biotechnol.* **2020**, *95*, 445–453. [CrossRef]
22. Gao, L.; Li, Y.; Wang, Z.; Sun, G.; Qi, X.; Mo, H. Physicochemical characteristics and functionality of tree peony (*Paeonia suffruticosa* Andr.) seed protein. *Food Chem.* **2018**, *240*, 980–988. [CrossRef]
23. Zhao, L.; Chen, Y.; Chen, Y.; Kong, X.; Hua, Y. Effects of pH on protein components of extracted oil bodies from diverse plant seeds and endogenous protease-induced oleosin hydrolysis. *Food Chem.* **2016**, *200*, 125–133. [CrossRef] [PubMed]
24. Zhao, L.; Chen, Y.; Cao, Y.; Kong, X.; Hua, Y. The integral and extrinsic bioactive proteins in the aqueous extracted soybean oil bodies. *J. Agric. Food Chem.* **2013**, *61*, 9727–9733. [CrossRef] [PubMed]
25. Yang, Y.; Leser, M.E.; Sher, A.A.; McClements, D.J. Formation and stability of emulsions using a natural small molecule surfactant: Quillaja saponin (Q-Naturale®). *Food Hydrocoll.* **2013**, *30*, 589–596. [CrossRef]
26. Li, Y.; Liu, X.; Liu, H.; Zhu, L. Interfacial adsorption behavior and interaction mechanism in saponin–protein composite systems: A review. *Food Hydrocoll.* **2023**, *136*, 108295. [CrossRef]
27. Ishii, T.; Matsumiya, K.; Nambu, Y.; Samoto, M.; Yanagisawa, M.; Matsumura, Y. Interfacial and emulsifying properties of crude and purified soybean oil bodies. *Food Struct.* **2017**, *12*, 64–72. [CrossRef]

28. Zhao, X.; Wang, K.; Dou, N.; Zhao, H.; Hou, D.; Wei, X.; Jiang, Z.; Hou, J. Oil body extraction from high-fat and high-protein soybeans by laccase cross-linked beet pectin: Physicochemical and oxidation properties. *J. Sci. Food Agric.* **2023**, *103*, 3550–3557. [CrossRef]
29. Ding, J.; Xu, Z.; Qi, B.; Jiang, L.; Sui, X. Physicochemical and oxidative stability of a soybean oleosome-based emulsion and its in vitro digestive fate as affected by (−)-epigallocatechin-3-gallate. *Food Funct.* **2018**, *9*, 6146–6154. [CrossRef]
30. Tippel, J.; Gies, K.; Harbaum-Piayda, B.; Steffen-Heins, A.; Drusch, S. Composition of Quillaja saponin extract affects lipid oxidation in oil-in-water emulsions. *Food Chem.* **2017**, *221*, 386–394. [CrossRef]

foods

MDPI

*Article*

# Profiling the Major Aroma-Active Compounds of Microwave-Dried Jujube Slices through Molecular Sensory Science Approaches

**Xinhuan Yan [1,2,†], Shaoxiang Pan [1,2,†], Xuemei Liu [1,2], Mengnan Tan [1,2], Xiaodong Zheng [1,2], Wenyu Du [1,2], Maoyu Wu [1,2] and Ye Song [1,2,*]**

1 Jinan Fruit Research Institute, All China Federation of Supply & Marketing Co-Operatives, Jinan 250014, China; yanjuan1206@126.com (X.Y.); panshaoxiang@126.com (S.P.); liuxm0218@163.com (X.L.); mntan28@163.com (M.T.); zhxd1106@163.com (X.Z.); dwy15169137686@163.com (W.D.); wmyu1972@hotmail.com (M.W.)
2 Shandong Province Fruit and Vegetable Storage and Processing Technology Innovation Center, Jinan 250014, China
* Correspondence: song___ye@163.com
† Xinhuan Yan and Shaoxiang Pan contributed equally to this work.

**Abstract:** To discriminate the aroma-active compounds in dried jujube slices through microwave-dried treatments and understand their sensory attributes, odor activity value (OAV) and detection frequency analysis (DFA) combined with sensory analysis and analyzed through partial least squares regression analysis (PLSR) were used collaboratively. A total of 21 major aromatic active compounds were identified, among which 4-hexanolide, 4-cyclopentene-1,3-dione, 5-methyl-2(5H)-furanone, 4-hydroxy-2,5-dimethyl-3(2H)furanone, 3,5-dihydroxy-2-methyl-4-pyrone were first confirmed as aromatic compounds of jujube. Sensory evaluation revealed that the major characteristic aromas of dried jujube slices were caramel flavor, roasted sweet flavor, and bitter and burnt flavors. The PLSR results showed that certain compounds were related to specific taste attributes. 2,3-butanedione and acetoin had a significant positive correlation with the roasted sweet attribute. On the other hand, γ-butyrolactone, 4-cyclopentene-1,3-dione, and 4-hydroxy-2,5-dimethyl-3(2H)furanone had a significant positive impact on the caramel attributes. For the bitter attribute, 2-acetylfuran and 5-methyl-2(5H)-furanone were positively correlated. Regarding the burnt flavor, 5-methyl-2-furancarboxaldehyde and 3,5-dihydroxy-2-methyl-4-pyrone were the most influential odor-active compounds. Finally, 2-furanmethanol and 2,3-dihydro-3,5-dihydroxy-6-methyl-4H-pyran-4-one were identified as the primary sources of the burnt and bitter flavors. Importantly, this work could provide a theoretical basis for aroma control during dried jujube slices processing.

**Keywords:** aroma-active compound; odor activity value; partial least squares regression; dried jujube slices

## 1. Introduction

The jujube (*Zizyphus jujuba* Mill., family Rhamnaceae), a traditional food in China, is attractive to consumers due to its rich nutrition, good taste, and potential health benefits [1]. The jujube was harvested from 26,000 hectares of land, and its production was 7.35 million tons in 2022 [2]. Since 2012, there has been an oversupply of jujube on the market because of the excessive production based on large-scale cultivation [3]. Notably, processing is an effective way to reduce post-harvest losses [4]. Among various processed jujube products, dried jujube slices are the most favorable, with the best taste, rich nutritional value, and easy to eat. Additionally, jujube slices provide good economic benefits and broad prospects for development [5]. Dried jujube slices are mostly prepared—including cleaning, coring, and slicing—from naturally dried jujube (about 30% moisture content). The main drying techniques include hot air, microwave, freeze, variable temperature, and pressure differential expansion drying [6]. Microwave drying is a widely applied method in the food industry, with easy operation, low cost, and a short drying cycle. Nowadays, researchers pay more attention to the technical energy consumption, appearance quality, and nutrient retention of different drying processes. Gao et al. [7] investigated the changes in sugars, organic acids, α-tocopherol, β-carotene, phenolic profiles, total phenolic content (TPC), and antioxidant capacity of jujube with four drying treatments (sun, oven, microwave, and freeze drying) [7]. Ji et al. [8] also compared the structural characterization and antioxidant activity of the polysaccharides of jujube (*Zizyphus jujuba* Mill., family Rhamnaceae) [8]. In contrast, research on the aroma of jujube slices has rarely been reported. Notably, the flavor is essential for the quality and acceptability of dried slices [9,10]. After microwave drying, the products could have a caramel and sweet roasted flavor similar to toast or coffee aroma under high temperatures and with enough oxygen. The unique caramel and roasted sweet flavor could benefit the products' good flavor [11]. However, unpleasant odor, such as burnt and bitter, could be obtained with an unsuitable drying procedure, which has a bad influence on jujube slices' quality. It is important to study the flavor components and their dynamic characteristics to optimize and control the process parameters during the processing of dried jujube slices.

Recently, some research regarding the aroma of dried jujube was carried out. Chen et al. [12] used headspace solid-phase microextraction gas chromatography–mass spectrometry (HS-SPME/GC-MS) and electronic nose (e-nose) to study a total of 51 aroma compounds from 10 different varieties of dried jujube [1,12]. Additionally, Song investigated the volatile compounds of Chinese jujubes with different drying methods based on metal oxide semiconductor (MOS) e-nose and flash GC e-nose [13]. The above reports illustrated that the drying procedures significantly influenced the variety and amount of volatile compounds and the flavor profile of dried jujube. However, there is a lack of theoretical basis for studying aroma changes in dried jujube during processing. Identifying effective aroma compounds offers a convenient approach to comprehending the sensory alterations, but their research and application in dried jujube slice products are relatively limited.

This study aimed to identify and measure the aroma-active compounds and their dynamics in dried jujube slices using GC-MS combined with GC-O. Additionally, we wanted to evaluate their contribution to the overall aroma by calculating the odor activity value (OAV) and detection frequency analysis (DFA) and to investigate the relationship between sensory properties and characteristic aromatic active compounds using multivariate analysis of partial least squares regression analysis (PLSR).

## 2. Materials and Methods

### 2.1. Raw Material

The jujube in the experiment was ordinary gray jujube purchased from Xinjiang jujube industry Co., Ltd. (Hetian, China) in August 2020 and transported to the laboratory within 72 h at room temperature. After arriving at the laboratory, the water and sugar contents were separately measured according to the method of Gao et al. [6], with 21.0% water content and 69.91% total sugar content.

### 2.2. Preparation of Microwave-Dried Jujube Slices

Before drying, approximately 10 kg of jujube was cut into slices, pitted, and baked in microwave equipment (SANLE, Nanjing, China) with 1 KW power. Seven kinds of jujube slices with different flavor characteristics were prepared by controlling the drying time (0 min, 1 min, 2 min, 2.5 min, 3 min, 4 min, and 5 min). Each drying experiment was carried out in triplicate. They were numbered as B1 (0 min), B2 (1 min), B3 (2 min), B4 (2.5 min), B5 (3 min), B6 (4 min), and B7 (5 min). Each sample was divided into two parts: approximately half was used for the sensory evaluation, and the other was grounded into powder by a food processor (FW100, Taisite Instrument Co., Ltd., Tianjin, China) for the aroma compounds' analyses. Subsequently, all samples were promptly packed in polyethylene bags under vacuum conditions to mitigate any potential alterations and stored at a temperature of −18 °C in a freezer until further analysis. Jujube powder sample (100 g), deionized water (320 g), 20% sodium chloride solution (48 g), and 1% sodium fluoride solution (32 g) were placed together in a pulper to make a puree. The purpose of adding 1% sodium fluoride solution is mainly to increase ionic strength, which is used to improve sensitivity [14]. Furthermore, the puree samples were immediately used for the analytical determinations with GC usage [14].

### 2.3. Analytical Determinations

#### 2.3.1. Extraction of Volatile Compounds through Drying Jujube with Headspace Solid-Phase Microextraction (HS-SPME)

Microextraction (HS-SPME) was used to extract volatile compounds from dried jujube slices. Non-polar poly dimethy Isiloxane (PDMS) fiber was preferred for the extraction of non-polar analytes. Mixed coating fibers, containing divinylbenzene (DVB) copolymers and carboxen (CAR), could increase retention capacity. PDMS/DVB and CAR/DVB were used for the extraction of low molecular weight volatile and polar analytes [15,16]. The 50/30/30 μm divinylbenzene/carboxen/polydimethylsiloxane (DVB/CAR/PDMS) coated fibers (Supelco, Bellefonte, PA, USA) were used to isolate the volatile compounds once the fibers were conditioned at 250 °C for 30 min. Before the formal experiment, we investigated the effects of the main parameters (including extraction time, sample amount, extraction temperature) on extraction efficiency. Three extraction times (20, 25, and 30 min), three different sample amounts (2, 5, and 8 g), and extraction temperatures (30, 40, and 50 °C) were assayed. The number of compounds and the total peak area were used to evaluate the extraction effect; then, the optimal optimization conditions were determined. For each sample, 5 g of jujube puree was added to a 20 mL headspace bottle (Supelco, Bellefonte, PA, USA) supplemented with 20 μL 2-octanol (32.88 μg/mL in methanol; Sigma-Aldrich, St. Louis, MO, USA) as an internal standard. The headspace bottles were placed in a multifunctional sampler (MPS 2XL; Gerstel GmbH & Co. KG, Mülheim an der Ruhr, Germany) and extracted for 25 min by vibrating at 40 °C. All experiments and sample measurements were carried out in triplicate, and the average values were recorded.

#### 2.3.2. GC-MS Analysis

A GC-MS apparatus (Thermo, Santa Clara, CA, USA) matched with a DB-WAX capillary column (30 mm × 0.25 mm × 0.25 μm) was utilized to isolate and identify volatile compounds in jujube slices. GC-MS (Thermo, Santa Clara, CA, USA) conditions were set as follows: kept at 50 °C for 7 min, ramped at 3 °C/min to 150 °C, increased at 10 °C/min to 250 °C, and maintained for 5 min. Helium, with a purity of 99.999%, served as the carrier gas at a flow rate of 1.5 mL/min in splitless GC inlet mode. MS fragmentation was conducted using electronic impact mode with an ionization energy of 70 eV and a source temperature of 250 °C. Moreover, the transmission line temperature was maintained at 250 °C. Full-scan mode was employed for acquisition, covering a mass range of 50–500 $m/z$.

#### 2.3.3. Qualitative and Semiquantitative Analyses

The volatiles were identified based on their mass spectra using the NIST10 library of the GC-MS data system, retention indices (RI) with reference values, and odor descriptions of authentic standards. Additionally, RI values were determined by employing n-alkanes, specifically C7–C30 from Sigma-Aldrich, as standards under the same instrument conditions. A compound was identified if the difference between the calculated and published RI values was less than 20.

With considerations given to cost effectiveness and practicality, an internal standard method was employed to quantify the identified volatiles. This approach aims to establish an efficient means of characterizing the behavior of aroma compounds. The concentration of each identified compound was calculated by comparing its peak area to the peak area of the internal standard (2-octanol). For ease of calculation, a calibration factor of 1.00 was utilized and determined using the following formula, according to the research results of Ref [17]:

$$m_s = \frac{m_i \times A_S}{A_i \times m_0} \times 1000 \quad (1)$$

In the formula, $m_s$ represents the concentrations of the identified volatiles in μg/kg; $m_i$ represents the weight of the internal standard in μg; $m_0$ represents the weight of the jujube slice used in grams; $A_S$ represents the peak area of the identified volatiles; and $A_i$ represents the peak area of the internal standard.

#### 2.3.4. Calculation of OAVs

The OAV is a metric used to assess the contribution of compounds to the overall aroma profile. It is calculated by dividing the concentration of compounds by their odor threshold in water. The odor threshold represents the minimum concentration humans can perceive in the compound. We referred to previous publications [18] to obtain the necessary data for calculating the OAV. Compounds with an OAV value equal to or greater than 1 are considered potential contributors to the aroma profile of the sample [19,20]. Importantly, these compounds are likely to noticeably impact the overall aroma perception.

#### 2.3.5. Gas Chromatography–Olfactometry (GC-O) Frequency Analysis

To characterize aroma-active compounds, an olfactory detector port (ODP 3; Gerstel GmbH & Co. KG) was employed in conjunction with a GC-MS instrument (Thermo, Santa Clara, CA, USA). This setup facilitated the differentiation of these compounds. After passing through the capillary column, the effluent was evenly split between the sniffing port and the MS detector. The transfer line leading to the GC-O sniffing port was maintained at 220 °C. To prevent nasal dryness, water was added to humidify the effluent, resulting in a flow rate of 60 mL/min. The remaining working conditions were consistent with those previously mentioned for the GC-MS analysis. Ten assessors, consisting of five males and five females, with over 300 h of technical experience, were selected. Prior to olfactory analysis, artificial odor solutions were used for odor identification. Assessors were instructed to record the samples' retention time and detection frequency (DF). An odorant with a DF (5) could be considered a potential contributor to the aroma profile [1,20].

#### 2.3.6. Sensory Evaluation

Sensory evaluations were conducted to describe the differences in dried jujube slices after undergoing various microwave treatments. The descriptors were selected from the words described by 10 professional reviewers, and words with 80% or more of the votes were counted as evaluation words. Subsequently, organoleptic characteristic descriptors were quantified using five sensory attributes (caramel flavor, roasted sweet flavor, bitter, burnt, and jujube-ID) [5]. The training primarily focused on sensory and semiquantitative descriptive analysis, following the ISO international standard (8586-1, 1993). During the 3-month training period, which consisted of 2 h per week, the panelists were trained to develop the ability to differentiate between different levels of aromas. The subsequent stage of training involved describing and discussing the aroma characteristics of dried jujube slices. The results were modified and annotated in the manuscript.

To evaluate the sensory characteristics of dried jujube slices treated with different microwave power, 5–6 slices of each sample were placed in a 50 mL plastic cup and covered. The cups were prepared 2 h before reaching headspace equilibrium at room temperature and randomly coded with a 3-digit number. Trained assessors used a 10-point scale with 1-unit increments from 0 to 10 to rate each descriptor, and there was a 3 min interval between each sample evaluation. Chinese panelists commonly use this scale due to its intuitive and easy-to-understand nature. Furthermore, the scale is broad enough to encompass the full sensory attribute intensities. It also has sufficient discrete points to distinguish subtle differences in intensity between samples [21,22]. The three independent QDA tests for each odor descriptor were averaged and plotted in a spider web chart.

*2.4. Statistical Analysis*

The experiments were conducted in triplicate, and the obtained data were analyzed using ANOVA. Duncan's multiple range test ($p < 0.05$) was employed to assess the differences. The statistical analysis was conducted using the SPSS software (SPSS Statistics 19.0; IBM Corporation, New York, NY, USA). Cluster analysis was accomplished using Rstudio (R-Tools Technology, Inc., Richmond Hill, ON, Canada).

The Unscrambler software version 9.7 (CAMO ASA, Oslo, Norway) was utilized to analyze the correlations between sensory attributes and volatile compounds and to generate the PLSR plot.

## 3. Results and Discussion

*3.1. Profile of Dried Jujube Slices*

Among the seven jujube slice samples, there were significant differences in the quantity and composition of the volatiles. A total of 83 volatile compounds were detected in the 7 samples, including 12 aldehydes, 21 esters, 18 ketones, 9 alcohols, 15 acids, 6 hydrocarbons, and 2 furans (Table 1). Most of the volatile compounds detected in this study agreed with previous research on dried jujube slices [23]. Compared with Ref [23], some new compounds were found, namely 4-hexanolide, 4-cyclopentene-1,3-dione, 5-methyl-2(5H)-furanone, 4-hydroxy-2,5-dimethyl-3(2H)furanone, and 3,5-dihydroxy-2-methyl-4-pyrone. Among the volatiles, acids accounted for the highest semiquantitative volatile portion (Figure 1), and 15 acids were identified in all samples. 4-Methylvaleric acid, trans-3-hexenoic acid, and decanoic acid were only detected in the B1 sample. The second highest semiquantitative volatile portion consisted of ketones in the B2–B7 samples (Figure 1). A total of 18 ketones were identified, the main components of which were furanone and pyranone. About 21 esters were identified during drying, regarded as one of the most important flavoring compounds in fresh jujube [24]. The most quantity-predominant compound in dried jujube slices was acetic acid (581.6 ug/kg in the B1 sample) (61 in Table 1). The acetic acid content in this study was consistent with previously reported quantification studies of jujube [25].

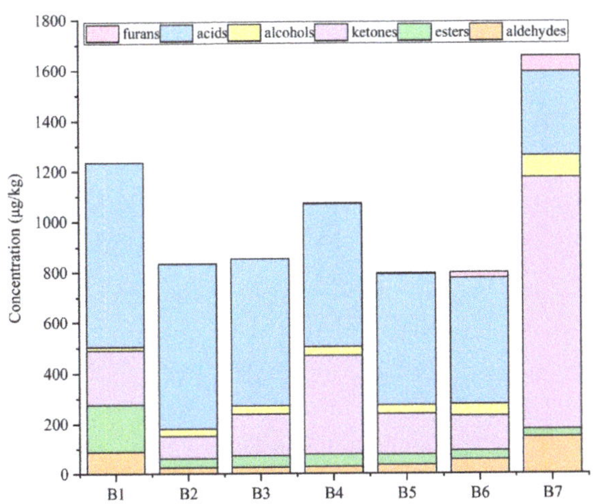


**Figure 1.** Effect of different microwave drying times on flavor and composition content of jujube slices: (B1) drying time 0 min; (B2) drying time 1 min; (B3) drying time 2 min; (B4) drying time 2.5 min; (B5) drying time 3 min; (B6) drying time 4 min; (B7) drying time 5 min.

**Table 1.** Volatile compounds' concentration (μg/kg) and their odor activity values (OAVs) in seven jujube slices.

| No | Compound Name [A] | CAS | RI [B] | LRI [C] | Identification [D] | Concentration (μg kg$^{-1}$) | | | | | | | OT [F] |
|---|---|---|---|---|---|---|---|---|---|---|---|---|---|
| | | | | | | B1 [I] | B2 | B3 | B4 | B5 | B6 | B7 | / |
| aldehydes | | | | | | | | | | | | | |
| 1 | 2-Butenal | 4170-30-3 | 1051 | - [E] | MS,RI | 42.9 ± 5.78 | 0 | 0 | 0 | 0 | 0 | 0 | / [H] |
| 2 | Hexanal | 66-25-1 | 1089 | 1081 | MS,RI | 15.39 ± 0.65 a [G] | 13.43 ± 0.23 b | 10.37 ± 0.31 c | 9.52 ± 1.12 d | 7.97 ± 0.74 e | 0 | 0 | 0.0036 |
| 3 | 2-Ethyl-2-butena | 19780-25-7 | 1158 | - | MS,RI | 3.64 ± 0.52 | 0 | 0 | 0 | 0 | 0 | 0 | / |
| 4 | (Z)-2-Heptenal | 57266-86-1 | 1271 | 1291 | MS,RI | 3.70 ± 0.47 | 0 | 0 | 0 | 0 | 0 | 0 | 0.01 |
| 5 | Nonanal | 124-19-6 | 1403 | 1392 | MS,RI | 8.38 ± 0.34 a | 7.34 ± 0.41 b | 5.32 ± 0.54 c | 4.23 ± 0.42 d | 3.11 ± 0.37 e | 2.26 ± 0.53 f | 1.02 ± 0.62 g | 0.0035 |
| 6 | (E,E)-2,4-Hexadienal | 142-83-6 | 1418 | 1403 | MS,RI | 1.17 ± 0.15 | 0 | 0 | 0 | 0 | 0 | 0 | 1.6 |
| 7 | Furfural | 98-01-1 | 1478 | 1466 | MS,RI | 0 | 0.93 ± 0.23 f | 5.38 ± 0.78 e | 8.34 ± 0.56 d | 17.04 ± 0.42 c | 38.97 ± 0.34 b | 82.5 ± 1.57 a | 0.008 |
| 8 | Decanal | 112-31-2 | 1508 | 1495 | MS,RI | 0 | 0.81 ± 0.11 a | 0.97 ± 0.09 a | 0.96 ± 0.07 a | 0.82 ± 0.15 a | 0.51 ± 0.21 b | 0 | 0.0009 |
| 9 | Benzaldehyde | 100-52-7 | 1541 | 1538 | MS,RI | 13.09 ± 1.12 | 0 | 0 | 0 | 0 | 0 | 0 | 0.05 |
| 10 | 5-Methyl-2-furanaldehyde | 620-02-0 | 1588 | 1570 | MS,RI | 0 | 1.56 ± 0.12 f | 3.03 ± 0.21 e | 3.87 ± 0.05 d | 5.34 ± 0.08 c | 13.82 ± 0.11 b | 23.19 ± 0.08 a | 0.005 |
| 11 | 5-Acetoxymethyl-2-furaldehyde | 10551-58-3 | 2203 | - | MS,RI | 0 | 0 | 0 | 0 | 0 | 0 | 2.33 ± 0.28 | / |
| 12 | 5-Hydroxymethyl-2-furaldehyde | 67-47-0 | 2514 | 2510 | MS,RI | 0 | 0 | 0.20 ± 0.04 d | 0.42 ± 0.05 c | 0.49 ± 0.09 c | 1.30 ± 0.25 b | 39.3 ± 2.27 a | 5 |
| esters | | | | | | | | | | | | | |
| 13 | Ethyl acetate | 108-05-4 | 989 | - | MS,RI | 0 | 14.4 ± 0.12 a | 6.76 ± 0.25 b | 4.20 ± 0.13 c | 4.76 ± 0.24 c | 3.58 ± 0.22 d | 1.02 ± 0.09 e | / |
| 14 | Ethyl valerate | 539-82-2 | 1141 | - | MS,RI | 12.2 ± 0.14 | 0 | 0 | 0 | 0 | 0 | 0 | 0.094 |
| 15 | Ethyl hexanoate | 123-66-0 | 1240 | 1238 | MS,RI | 139.24 ± 5.67 a | 4.04 ± 0.21 b | 3.49 ± 0.15 b | 3.50 ± 0.22 b | 2.11 ± 0.13 c | 0 | 0 | 0.0005 |
| 16 | Ethyl heptanoate | 106-30-9 | 1340 | - | MS,RI | 10.04 ± 0.52 | 0 | 0 | 0 | 0 | 0 | 0 | 0.17 |
| 17 | 2-Hexenoic acid ethyl ester | 1552-67-6 | 1353 | - | MS,RI | 3.82 ± 0.21 | 0 | 0 | 0 | 0 | 0 | 0 | / |
| 18 | 2,3-Dihydro-5-methyl-2-furanone | 591-12-8 | 1418 | - | MS,RI | 0 | 0.53 ± 0.04 c | 0.76 ± 0.06 bc | 0.84 ± 0.06 b | 0.86 ± 0.09 b | 0.89 ± 0.12 b | 8.92 ± 0.82 a | / |
| 19 | Ethyl caprylate | 106-32-1 | 1441 | 1466 | MS,RI | 6.53 ± 0.32 | 0 | 0 | 0 | 0 | 0 | 0 | 0.0001 |
| 20 | (E)-9-Tetradecen-1-olacetate | 23192-82-7 | 1478 | - | MS,RI | 0.91 ± 0.08 | 0 | 0 | 0 | 0 | 0 | 0 | / |
| 21 | Formic acid furfuryl ester | 13493-97-5 | 1507 | 1481 | MS,RI | 0 | 0 | 0 | 0 | 0 | 0 | 2.44 ± 0.12 | / |
| 22 | Methyl decanoate | 110-42-9 | 1601 | 1590 | MS,RI | 0.44 ± 0.08 | 0 | 0 | 0 | 0 | 0 | 0 | 1 |
| 23 | 4-Hydroxy-2-methylbutanoic acid lactone | 1679-47-6 | 1610 | - | MS,RI | 0.76 ± 0.11 b | 1.21 ± 0.14 a | 0.95 ± 0.08 b | 0.92 ± 0.06 b | 0.89 ± 0.06 b | 0.86 ± 0.10 b | 0 | / |
| 24 | γ-Valerolactone | 108-29-2 | 1630 | - | MS,RI | 0 | 0.56 ± 0.05 a | 0.50 ± 0.04 ab | 0.44 ± 0.04 b | 0.41 ± 0.03 b | 0.33 ± 0.04 c | 0.30 ± 0.03 c | 100 |
| 25 | Ethyl caprate | 110-38-3 | 1644 | - | MS,RI | 4.25 ± 0.27 | 0 | 0 | 0 | 0 | 0 | 0 | 0.02 |
| 26 | γ-Butyrolactone | 96-48-0 | 1650 | - | MS,RI | 1.34 ± 0.25 g | 10.05 ± 0.42 e | 28.80 ± 0.52 c | 36.65 ± 0.37 a | 30.24 ± 0.48 b | 25.64 ± 0.62 d | 2.55 ± 0.44 f | 0.025 |
| 27 | 4-Hexanolide | 695-06-7 | 1722 | - | MS,RI | 4.04 ± 0.27 a | 3.71 ± 0.36 a | 3.50 ± 0.52 ab | 2.97 ± 0.21 b | 2.53 ± 0.35 b | 2.43 ± 0.22 b | 2.37 ± 0.19 b | 8 |
| 28 | δ-Hexanolactone | 823-22-3 | 1815 | - | MS,RI | 0.76 ± 0.05 d | 1.52 ± 0.12 a | 1.51 ± 0.11 a | 1.33 ± 0.13 ab | 1.24 ± 0.08 b | 1.20 ± 0.07 b | 0.93 ± 0.05 c | 230 |
| 29 | γ-Heptanolactone | 105-21-5 | 1825 | - | MS,RI | 0.60 ± 0.05 | 0 | 0 | 0 | 0 | 0 | 0 | 0.52 |
| 30 | Ethyl laurate | 106-33-2 | 1849 | - | MS,RI | 0.73 ± 0.04 | 0 | 0 | 0 | 0 | 0 | 0 | 3.5 |
| 31 | γ-Octanoic lactone | 104-50-7 | 1935 | 1883 | MS,RI | 0.55 ± 0.06 b | 0.52 ± 0.05 b | 0.69 ± 0.04 a | 0.59 ± 0.04 ab | 0.55 ± 0.06 b | 0.67 ± 0.06 a | 0.56 ± 0.03 b | 0.095 |
| 32 | Methyl pyruvate | 600-22-6 | 2357 | - | MS,RI | 0 | 0 | 0 | 0 | 0 | 0 | 10.39 ± 1.24 | / |
| 33 | 3,4-Dihydroxybutanoic acid gamma-lactone | 5469-16-9 | 2618 | - | MS,RI | 0 | 0 | 0 | 0 | 0 | 0 | 0.92 ± 0.08 | / |
| ketones | | | | | | | | | | | | | |
| 34 | 2,3-Butanedione | 431-03-8 | 1022 | 981 | MS,RI | 0 | 12.00 ± 2.58 d | 29.00 ± 3.12 b | 38.00 ± 1.11 a | 22.00 ± 0.85 c | 7.00 ± 0.41 e | 2.00 ± 0.34 f | 0.003 |
| 35 | Acetoin | 513-86-0 | 1294 | 1292 | MS,RI | 0 | 60.28 ± 4.28 d | 102.09 ± 3.78 b | 128.23 ± 4.12 a | 87.45 ± 3.56 c | 42.16 ± 2.87 e | 8.25 ± 1.56 f | 0.04 |
| 36 | 6-Methyl-5-hepten-2-one | 110-93-0 | 1347 | 1339 | MS,RI | 0 | 1.49 ± 0.22 b | 1.74 ± 0.15 b | 1.81 ± 0.2 ab | 1.84 ± 0.34 ab | 2.36 ± 0.45 a | 2.50 ± 0.52 a | 0.1 |
| 37 | 3-Acetoxy-2-butanone | 4906-24-5 | 1390 | - | MS,RI | 3.18 ± 0.12 d | 6.46 ± 0.23 a | 5.59 ± 0.18 b | 4.41 ± 0.55 c | 4.38 ± 0.42 c | 3.64 ± 0.43 d | 2.08 ± 0.15 e | / |
| 38 | 4-Cyclopentene-1,3-dione | 930-60-9 | 1602 | - | MS,RI | 2.12 ± 0.15 g | 8.56 ± 0.05 e | 18.24 ± 0.07 c | 28.37 ± 0.07 a | 20.54 ± 0.06 b | 12.65 ± 0.08 d | 5.12 ± 0.14 f | 0.02 |

**Table 1.** *Cont.*

| No | Compound Name [A] | CAS | RI [B] | LRI [C] | Identification [D] | Concentration (μg $kg^{-1}$) | | | | | | | OT [F] |
|---|---|---|---|---|---|---|---|---|---|---|---|---|---|
| | | | | | | B1 [I] | B2 | B3 | B4 | B5 | B6 | B7 | / |
| 39 | 5,5-Dimethylfuran-2(5H)-one | 20019-64-1 | 1626 | - | MS,RI | 0 | 0.42 ± 0.05 a | 0.44 ± 0.04 a | 0.45 ± 0.06 a | 0.46 ± 0.02 a | 0.47 ± 0.04 a | 0.50 ± 0.03 a | / |
| 40 | 5-Methyl-2(5H)-furanone | 591-11-7 | 1698 | - | MS,RI | 0 | 0 | 0.01 ± 0.002 d | 0.97 ± 0.05 c | 1.26 ± 0.04 bc | 6.99 ± 0.05 b | 15.02 ± 0.15 a | 0.002 |
| 41 | 2(5H)-Furanone | 497-23-4 | 1776 | - | MS,RI | 0 | 0 | 0.24 ± 0.03 d | 0.25 ± 0.04 cd | 0.32 ± 0.04 c | 0.48 ± 0.05 b | 2.66 ± 0.21 a | / |
| 42 | 1,2-Cyclopentanedione | 3008-40-0 | 1783 | - | MS,RI | 0 | 0 | 0 | 0 | 0 | 0 | 1.03 ± 0.11 | / |
| 43 | 3-Methyl-1,2-cyclopentanedione | 765-70-8 | 1842 | - | MS,RI | 0 | 0.61 ± 0.08 c | 0.95 ± 0.07 b | 0.98 ± 0.12 b | 1.00 ± 0.09 b | 1.07 ± 0.13 ab | 1.42 ± 0.24 a | 0.01 |
| 44 | Furyl hydroxymethyl ketone | 17678-19-2 | 2023 | - | MS,RI | 0 | 0 | 0 | 0 | 0 | 0 | 1.03 ± 0.12 | 1 |
| 45 | 4-Hydroxy-2,5-dimethyl-3(2H)furanone | 3658-77-3 | 2038 | 2002 | MS,RI | 0 | 0.41 ± 0.09 c | 0.96 ± 0.24 b | 1.75 ± 0.42 a | 0.79 ± 0.15 b | 0.46 ± 0.07 c | 0.38 ± 0.04 cd | 0.001 |
| 46 | 5-Acetyldihydrofuran-2(3H)-one | 29393-32-6 | 2080 | - | MS,RI | 0 | 0 | 0 | 0 | 0 | 0.42 ± 0.05 b | 2.13 ± 0.12 a | / |
| 47 | 2,5-Hexanedione | 110-13-4 | 2120 | - | MS,RI | 0 | 0 | 0 | 0 | 0 | 0 | 0.60 ± 0.05 | / |
| 48 | 2-Hydroxy-gamma-butyrolactone | 19444-84-9 | 2188 | - | MS,RI | 0 | 0 | 0 | 0 | 0 | 0.43 ± 0.03 b | 3.99 ± 0.11a | / |
| 49 | 2,3-Dihydro-3,5-dihydroxy-6-methyl-4(H)-pyran-4-one | 28564-83-2 | 2286 | 2266 | MS,RI | 0 | 1.13 ± 0.12 f | 5.92 ± 0.24 e | 11.57 ± 0.21 d | 16.50 ± 0.32 c | 53.04 ± 0.27 b | 106.79 ± 1.24 a | 0.02 |
| 50 | 3,5-Dihydroxy-2-methyl-4-pyrone | 1073-96-7 | 2314 | - | MS,RI | 0 | 0 | 0.54 | 1.12 | 3.56 | 7.35 | 13.12 ± 0.08 | 0.01 |
| 51 | Methyl pyruvate | 600-22-6 | 2357 | - | MS,RI | 0 | 0 | 0 | 0 | 0 | 0 | 10.39 ± 0.85 | / |
| alcohols | | | | | | | | | | | | | |
| 52 | 1-Octen-3-ol | 3391-86-4 | 1453 | 1441 | MS,RI | 5.69 ± 0.52 a | 4.87 ± 0.37 bc | 4.92 ± 0.35 b | 4.69 ± 0.41 bc | 4.52 ± 0.25 c | 4.23 ± 0.36 c | 0 | 0.001 |
| 53 | 2-Ethylhexanol | 104-76-7 | 1493 | - | MS,RI | 6.87 ± 0.24 a | 4.58 ± 0.35 c | 4.73 ± 0.24 c | 4.84 ± 0.52 bc | 4.92 ± 0.32 bc | 5.35 ± 0.29 b | 7.20 ± 0.54 a | / |
| 54 | (S,S)-2,3-Butanediol | 19132-06-0 | 1543 | - | MS,RI | 0 | 13.30 ± 0.55 a | 12.42 ± 0.47 b | 12.01 ± 0.85 bc | 11.72 ± 0.65 c | 10.59 ± 0.74 c | 0 | 400 |
| 55 | Propylene glycol | 57-55-6 | 1595 | - | MS,RI | 0.83 ± 0.08 c | 1.66 ± 0.23 a | 1.47 ± 0.21 ab | 1.45 ± 0.16 ab | 1.35 ± 0.12 b | 1.31 ± 0.17 b | 1.24 ± 016 b | 1400 |
| 56 | 4-Methyl-5-decanol | 213547-15-0 | 1659 | - | MS,RI | 0.38 ± 0.05 | 0 | 0 | 0 | 0 | 0 | 0 | / |
| 57 | 2-Furanmethanol | 98-00-0 | 1669 | - | MS,RI | 0 | 1.49 ± 0.18 f | 5.57 ± 0.24 e | 7.23 ± 0.32 d | 8.78 ± 0.34 c | 20.44 ± 1.24 b | 73.57 ± 3.25 a | 0.3 |
| 58 | 5-Methyl-2-furanMethanol | 3857-25-8 | 1728 | - | MS,RI | 0 | 0 | 0.64 ± 0.07 c | 0.75 ± 0.09 c | 0.78 ± 0.08 c | 1.64 ± 0.15 b | 2.57 ± 0.24 a | 0.3 |
| 59 | Benzyl alcohol | 100-51-6 | 1888 | - | MS,RI | 0 | 0.23 ± 0.05 a | 0.23 ± 0.03 a | 0.25 ± 0.02 a | 0.22 ± 0.03 a | 0.27 ± 0.04 a | 0.23 ± 0.03 a | 5.5 |
| 60 | (R)-(-)-3-Methyl-2-butanol | 1572-93-6 | 1982 | - | MS,RI | 0 | 2.27 ± 0.24 b | 2.58 ± 0.31 ab | 2.68 ± 0.26 a | 2.74 ± 0.27 a | 2.56 ± 0.18 ab | 2.10 ± 0.12 b | / |
| acids | | | | | | | | | | | | | |
| 61 | Acetic acid | 64-19-7 | 1459 | 1442 | MS,RI | 581.59 ± 2.24 c | 509.50 ± 2.25 a | 456.68 ± 3.57 b | 451.40 ± 4.28 b | 414.41 ± 4.35 c | 401.04 ± 2.15 d | 254.97 ± 1.57 e | 0.1 |
| 62 | Propanoic acid | 79-09-4 | 1550 | 1531 | MS,RI | 11.51 ± 0.25 d | 14.25 ± 0.32 a | 12.54 ± 0.18 b | 12.06 ± 0.15 c | 11.21 ± 0.21 de | 11.14 ± 0.17 e | 10.60 ± 0.27 f | 3 |
| 63 | Butanoic acid | 107-92-6 | 1639 | - | MS,RI | 31.77 ± 0.25 | 26.08 ± 0.32 | 24.93 ± 0.21 | 21.91 ± 0.18 | 20.40 ± 0.12 d | 17.73 ± 0.08 | 16.04 ± 0.11 g | 0.015 |
| 64 | Isovaleric acid | 503-74-2 | 1680 | 1626 | MS,RI | 24.87 ± 0.52 a | 22.85 ± 0.47 b | 17.17 ± 0.32 c | 16.83 ± 0.25 cd | 14.55 ± 0.42 d | 13.47 ± 0.38 e | 8.36 ± 0.51 f | 0.1 |
| 65 | Pentanoic acid | 109-52-4 | 1749 | 1714 | MS,RI | 10.51 ± 0.25 c | 12.81 ± 0.35 a | 11.04 ± 0.19 b | 10.80 ± 0.42 bc | 8.91 ± 0.32 d | 8.34 ± 0.26 d | 6.10 ± 0.21 e | 0.5 |
| 66 | Isocrotonic acid | 503-64-0 | 1791 | - | MS,RI | 1.86 ± 0.12 d | 2.80 ± 0.08 a | 2.39 ± 0.07 b | 2.43 ± 0.11 b | 2.33 ± 0.08 b | 2.22 ± 0.07 bc | 2.15 ± 0.08 c | / |
| 67 | Hexanoic acid | 142-62-1 | 1855 | 1855 | MS,RI | 55.27 ± 0.85 a | 55.64 ± 1.21 a | 49.09 ± 0.54 b | 41.68 ± 0.36 c | 38.78 ± 0.41 d | 37.76 ± 0.52 e | 26.99 ± 0.44 f | 0.2 |
| 68 | 4-Methylvaleric acid | 646-07-1 | 1878 | - | MS,RI | 0.29 ± 0.02 | 0 | 0 | 0 | 0 | 0 | 0 | / |
| 69 | trans-3-Hexenoic acid | 1577-18-0 | 1954 | 1958 | MS,RI | 0.25 ± 0.03 | 0 | 0 | 0 | 0 | 0 | 0 | 1.3 |
| 70 | Heptanoic acid | 111-14-8 | 1960 | 1948 | MS,RI | 4.74 ± 0.24 a | 4.51 ± 0.32 a | 4.30 ± 0.38 ab | 3.97 ± 0.42 b | 3.92 ± 0.35 b | 3.58 ± 0.37 b | 2.44 ± 0.21 c | 0.1 |
| 71 | trans-2-Hexenoic acid | 1191-04-4 | 1982 | - | MS,RI | 1.51 ± 0.11 | 0 | 0 | 0 | 0 | 0 | 0 | / |
| 72 | Octanoic acid | 124-07-2 | 2066 | - | MS,RI | 2.94 ± 0.24 a | 3.09 ± 0.26 a | 2.80 ± 0.13 a | 2.73 ± 0.32 ab | 2.64 ± 0.21 b | 2.36 ± 0.18 b | 2.12 ± 0.15 c | 0.5 |
| 73 | trans-2-Undecenoic acid | 15790-94-0 | 2091 | - | MS,RI | 0.71 ± 0.05 b | 0.88 ± 0.06 a | 0.85 ± 0.05 a | 0.79 ± 0.03 ab | 0.77 ± 0.04 b | 0.61 ± 0.03 c | 0 | / |
| 74 | Nonanoic acid | 112-05-0 | 2177 | 2177 | MS,RI | 0.74 ± 0.07 ab | 0.95 ± 0.08 a | 0.93 ± 0.06 a | 0.86 ± 0.08 a | 0.76 ± 0.07 ab | 0.70 ± 0.06 b | 0.43 ± 0.05 c | 0.0035 |
| 75 | Decanoic acid | 334-48-5 | 2288 | - | MS,RI | 2.16 ± 0.15 | 0 | 0 | 0 | 0 | 0 | 0 | 0.5 |

**Table 1.** *Cont.*

| No | Compound Name [A] | CAS | RI [B] | LRI [C] | Identification [D] | Concentration ($\mu g\ kg^{-1}$) | | | | | | | OT [F] |
|---|---|---|---|---|---|---|---|---|---|---|---|---|---|
| | | | | | | B1 [I] | B2 | B3 | B4 | B5 | B6 | B7 | / |
| hydrocarbons | | | | | | | | | | | | | |
| 76 | Undecane | 1120-21-4 | 1104 | - | MS,RI | 0 | 0 | 0 | 0 | 0 | 0.88 ± 0.07 b | 8.25 ± 0.12 a | / |
| 77 | Dodecane | 112-40-3 | 1203 | 1200 | MS,RI | 0 | 14.01 ± 0.12 e | 14.63 ± 0.08 d | 14.16 ± 0.11 e | 15.82 ± 0.12 c | 20.71 ± 0.15 b | 32.30 ± 0.32 a | / |
| 78 | 2,5,6-Trimethyloctane | 62016-14-2 | 1209 | - | MS,RI | 1.61 ± 0.08 | 0 | 0 | 0 | 0 | 0 | 0 | / |
| 79 | Styrene | 100-42-5 | 1267 | 1250 | MS,RI | 0 | 3.39 ± 0.28 d | 4.19 ± 0.23 c | 4.29 ± 0.35 bc | 4.59 ± 0.24 b | 5.08 ± 0.41 a | 5.45 ± 0.32 a | 0.022 |
| 80 | Tridecane | 629-50-5 | 1302 | 1300 | MS,RI | 0 | 4.66 ± 0.33 d | 6.75 ± 0.41 c | 8.13 ± 0.38 b | 8.66 ± 0.35 b | 8.96 ± 0.54 b | 22.73 ± 1.12 a | / |
| 81 | Nonane | 111-84-2 | 1310 | 900 | MS,RI | 1.61 ± 0.23 e | 1.14 ± 0.21 e | 4.19 ± 0.52 cd | 4.84 ± 0.38 c | 3.52 ± 0.32 d | 7.99 ± 0.56 b | 24.06 ± 1.21 a | / |
| furans | | | | | | | | | | | | | |
| 82 | 2-Acetylfuran | 1192-62-7 | 1518 | - | MS,RI | 0 | 0.45 ± 0.09 e | 1.06 ± 0.12 d | 3.21 ± 0.26 c | 3.67 ± 0.21 c | 22.40 ± 0.12 b | 62.50 ± 0.41 a | 0.01 |
| 83 | Furan-2,5-dicarbaldehyde | 823-82-5 | 1999 | - | MS,RI | 0 | 0 | 0 | 0 | 0 | 0 | 0.77 ± 0.08 | 100 |

[A] The aroma compounds identified on the TG-WAXMS column. [B] The retention index of volatile compounds on the TG-WAXMS column. [C] The retention index of volatile compounds from published literature and online library. [D] RI: retention index; MS: mass spectrometry [26]. [E] The aroma compounds not identified in published literature and online library [27]. [F] The threshold of volatile compounds in water referred to in the book (Rychlik M, Schieberle P, Grosch W. *Compilation of odor thresholds, odor qualities and retention indices of key food odorants* [M]. Germany: Deutche Forschungsanstalt fur Lebensmittelchemie: Garching, 1998). [G] Values with different superscript Roman letters (a–g) in the same row are significantly different according to the Duncan test ($p < 0.05$). [H] The threshold of volatile compounds not found. [I] (B1) Microwave drying time 0 min; (B2) Microwave drying time 1 min; (B3) Microwave drying time 2 min; (B4) Microwave drying time 2.5 min; (B5) Microwave drying time 3 min; (B6) Microwave drying time 4 min; (B7) Microwave drying time 5 min.

### 3.2. Identification of Aroma-Active Compounds by OAV

The OAVs have been widely used to assess the aroma potency of foods by considering the balance between the food matrix and the surrounding air, as reported in previous studies [28,29]. In this study, the OAVs of 83 volatile compounds with documented odor threshold values, as shown in Table 1, were calculated to evaluate their contribution to the overall aroma of dried jujube slices [30]. Thirty-one volatile compounds were excluded from the analysis due to insufficient odor threshold data. Among the remaining compounds, 17 were identified as aroma-active based on their OAV values, with a relatively high contribution to the overall aroma. The OAV contribution was dominated by esters (94.03%) in the B1 sample, with hexyl acetate, which had a strong fruity aroma, having the highest OAV of 278.48. Ketones (33.88–42.86%) dominated their OAV contribution in B3–B6 samples, three of which, with the strong roasted sweet-like 2,3-Butanedione aroma, had the highest OAV of 2.3–12.7. Furthermore, furanones and pyranones, typically produced by lipid peroxidation and Maillard reactions of carbonyl and amino compounds, can significantly impact the aroma profile of food [31]. However, in identifying aroma-active compounds, it is also essential to consider detection frequency analysis findings.

### 3.3. Identification of Aroma-Active Compounds by DFA

DFA is a common method for identifying aroma compounds, as it allows for characterizing compounds, which contribute to the overall aroma perception [32]. To be considered an aroma-active compound, it must be detected by at least 5 out of the 10 panelists in the discrimination test.

Lina Wang [33] used "roast", "sweet", "green", "sour", "fruity", and "mellow" attributes to describe the odor of jujubes. Moreover, bitter and jujube-ID were evaluated as important attributes in analyzing the flavor of Spanish jujube (*Zizyphus jujuba* Mill., family Rhamnaceae) fruits [34]. Notably, different substances are present in different odors. A total of 15 aromatic aroma compounds were identified by matching the LRI and mass spectrum. These compounds were then categorized into five groups according to their odor characteristics. These were roasted sweet odorants (34 and 35 in Table 1), caramel-like odorants (26 and 38 in Table 1), burnt-like odorants (10, 12, 49, 50, and 57 in Table 1), bitter-like odorants (40, 49, 50, and 82 in Table 1), and jujube-ID odorants (13, 27, 52, and 67 in Table 1). The jujube-ID odorants consisted of mushroom, smelly, sour, and sweet substances.

### 3.4. Comparison of DFA and OAV Aroma-Active Compounds' Identification

Eleven volatile compounds were identified as potential aroma-active compounds in dried jujube slices based on the joint analysis of OAV and DFA, such as 4-hexanolide, 5-methyl-2(5H)-furanone, 3,5-dihydroxy-2-methyl-4-pyrone, 4-hydroxy-2,5-dimethyl-3(2H)furanone, 4-cyclopentene-1,3-dione, and so on. Some compounds with high DF could not be detected by OAV, such as 2,3-butanedione and acetoin in the B7 sample, which were significantly and positively correlated with roasted sweet attributes. Moreover, some compounds with high OAV could not be recognized by DFA, such as butanoic acid with descriptors of smelly. In food matrices, the release of aroma compounds can be influenced by interactions between the volatiles and the components of the food [1]. Thus, we should use both methods to identify aromatic active compounds. The results of OAV were in good agreement with DFA in identifying the major contributors to the aroma of dried jujube slices. Furthermore, 21 compounds were recognized as significant aroma-active compounds in this study. The cluster heat map of the 21 compounds of the samples under different treatment conditions is shown in Figure 2.

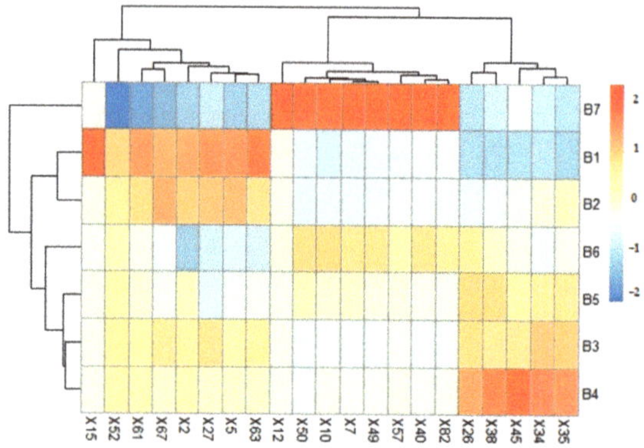


**Figure 2.** Hierarchical clustering of the 21 major aroma-active compounds in samples with different microwave drying times: (B1) drying time 0 min; (B2) drying time 1 min; (B3) drying time 2 min; (B4) drying time 2.5 min; (B5) drying time 3 min; (B6) drying time 4 min; (B7) drying time 5 min.

### *3.5. Effect of Different Microwave Drying Times on the Major Aroma-Active Compounds in Jujube*

To characterize the aroma profile of jujube slices dried with microwave treatments, major aroma-active compounds (21 in total) were identified through OAV and DFA, and the results are presented in Table 2. Hierarchical clustering, including heat maps (Figure 2), was performed based on the changes in the identified aroma compounds to illustrate the aroma profile of the seven jujube slice samples with different processing methods. Cluster analysis indicated that seven samples were distinguished into three classes (B1 and B2, B3 to B6, B7). Moreover, the changes in major aroma-active compounds and the widening disparities from B1 to B7 with the extension of drying time indicated that the characteristic aroma changed the processing.

Esters are regarded as one of jujube's most important flavoring compounds [24]. Among the three aroma-active esters, ethyl hexanoate (fruity) was found to decline significantly with the increase in drying time, and it was speculated that an increase in the heat treatment time could promote the degradation of ethyl hexanoate, which agrees with the findings of previous studies.

5-Acetyldihydrofuran-2(3H)-one (sweet) was first discovered in jujube, and the variation tendency agreed with ethyl hexanoate. As for the microwave treatments, the concentrations of $\gamma$-butyrolactone tended to rise first and then decrease along with the increase in drying time. One possible explanation for this effect is the inactivation of ester synthase and the Maillard reaction of valine and isoleucine, which only occurs within specific drying time ranges. However, additional research is needed to confirm this proposed explanation.

Five aroma-active aldehydes were identified in jujube slices, including hexanal and nonanal. These compounds are commonly associated with "green", "cut grass", "fat", and "citrus" notes, which were also noted by panelists in the sensory descriptions of jujube slices and are typical of many other jujube varieties [3,35,36]. Hexanal and nonanal showed a decreasing tendency in response to drying time, possibly due to thermal degradation and volatilization loss. Furfural, 5-methyl-2-furanaldehyde, and 5-hydroxymethyl-2-furaldehyde were not detected in the fresh jujube and significantly increased ($p < 0.05$) with the increase in drying time. It is well known that furfural and furan aldehydes are formed by the Maillard reaction of pentose or hexose [37,38], contributing to burnt attributes. Based on this theory, the laws of generation and change of these three aldehydes can be well explained.

Ketones and furans were the most diverse components among the major aroma-active compounds of jujube slices. In contrast, seven aroma-impacting ketones and one furan were found in the jujube samples. They were 2,3-butanedione (roasted sweet), acetoin (roasted sweet), 4-hydroxy-2,5-dimethyl-3(2H)furanone (caramel), 4-cyclopentene-

1,3-dione (caramel), 5-methyl-2(5H)-furanone (bitter), 2,3-dihydro-3,5-dihydroxy-6-methyl-4(H)-pyran-4-one (burnt, bitter), 2-acetylfuran (bitter), 3,5-dihydroxy-2-methyl-4-pyrone (burnt), which were formed during heating, except 4-cyclopentene-1,3-dione. The results of olfactometric analysis and compound flavor description showed that ketones and furans were important parts of the flavor of dried jujube slices. Studies have also shown that the formation of ketones and furans was related to the Maillard reaction, Strecker degradation reaction, and secondary reaction of fat thermal oxidation and thermal degradation [24]. The concentrations of the three ketones 4-hydroxy-2,5-dimethyl-3(2H)furanone, 4-cyclopentene-1,3-dione, and acetoin(2,3-butanedione) decreased, starting with sample B5. This may be because the intermediate ketones continue to react with amino acids to produce melaniginoids with the intensification of the Maillard reaction. However, the pattern of change in ketones requires further study and discussion.

**Table 2.** The major odor-active compounds identified by GC-O in seven jujube slices with the odor activity value (OAV) method and detection frequency analysis (DFA).

| No | Compound Name | CAS | OAV [A] | | | | | | | DF [B] | | | | | | | Odor Quality [C] |
|---|---|---|---|---|---|---|---|---|---|---|---|---|---|---|---|---|---|
| | | | B1 | B2 | B3 | B4 | B5 | B6 | B7 | B1 | B2 | B3 | B4 | B5 | B6 | B7 | |
| aldehydes | | | | | | | | | | | | | | | | | |
| 2 | Hexanal | 66-25-1 | 4.3 | 3.7 | 2.9 | 2.6 | 2.2 | <1 | <1 | 1 | 0 [D] | 0 | 0 | 0 | 0 | 0 | green |
| 5 | Nonanal | 124-19-6 | 2.4 | 2.1 | 1.5 | 1.2 | 0.9 | 0.6 | 0.3 | 1 | 0 | 0 | 0 | 0 | 0 | 0 | fatty |
| 7 | Furfural | 98-01-1 | <1 | <1 | <1 | 1 | 2.1 | 4.9 | 10.3 | 0 | 0 | 0 | 0 | 0 | 0 | 3 | burnt |
| 10 | 5-Methyl-2-furanaldehyde | 620-02-0 | <1 | <1 | <1 | <1 | 1.07 | 2.76 | 4.64 | 0 | 0 | 0 | 0 | 1 | 5 | 7 | burnt |
| 12 | 5-Hydroxymethyl-2-furaldehyde | 67-47-0 | <1 | <1 | <1 | <1 | <1 | <1 | <1 | 0 | 0 | 0 | 0 | 1 | 4 | 6 | burnt |
| esters | | | | | | | | | | | | | | | | | |
| 15 | Ethyl hexanoate | 123-66-0 | 278.5 | 8.1 | 7 | 7 | 4.2 | <1 | <1 | 3 | 2 | 1 | 1 | 0 | 0 | 0 | fruity |
| 26 | γ-Butyrolactone | 96-48-0 | <1 | <1 | 1.15 | 1.47 | 1.21 | 1.03 | <1 | 5 | 5 | 6 | 6 | 6 | 5 | 5 | caramel |
| 27 | 4-Hexanolide | 695-06-7 | <1 | <1 | <1 | <1 | <1 | <1 | <1 | 0 | 0 | 0 | 0 | 3 | 6 | 7 | sweet |
| ketones | | | | | | | | | | | | | | | | | |
| 34 | 2,3-Butanedione | 431-03-8 | <1 | 4 | 9.7 | 12.7 | 7.3 | 2.3 | <1 | 9 | 10 | 10 | 10 | 10 | 10 | 10 | roasted sweet |
| 35 | Acetoin | 513-86-0 | <1 | 1.5 | 2.6 | 3.2 | 2.2 | 1 | <1 | 10 | 10 | 10 | 10 | 10 | 10 | 10 | roasted sweet |
| 38 | 4-Cyclopentene-1,3-dione | 930-60-9 | <1 | <1 | <1 | 1.42 | 1.03 | <1 | <1 | 0 | 0 | 2 | 6 | 3 | 2 | 2 | caramel |
| 40 | 5-Methyl-2(5H)-furanone | 591-11-7 | 0 | 0 | 0.005 | 0.49 | 0.63 | 3.5 | 7.51 | 0 | 0 | 0 | 0 | 3 | 6 | 9 | bitter |
| 45 | 4-Hydroxy-2,5-dimethyl-3(2H)furanone | 3658-77-3 | <1 | <1 | <1 | 1.75 | <1 | <1 | <1 | 0 | 0 | 1 | 3 | 2 | 1 | 1 | caramel |
| 49 | 2,3-Dihydro-3,5-dihydroxy-6-methyl-4(H)-pyran-4-one | 28564-83-2 | 0 | 0.06 | 0.3 | 0.58 | 0.83 | 2.65 | 5.34 | 0 | 0 | 1 | 3 | 7 | 8 | 8 | burnt, bitter |
| 50 | 3,5-Dihydroxy-2-methyl-4-pyrone | 1073-96-7 | <1 | <1 | <1 | <1 | <1 | <1 | 1.31 | 0 | 0 | 0 | 0 | 1 | 5 | 6 | burnt, bitter |
| alcohols | | | | | | | | | | | | | | | | | |
| 52 | 1-Octen-3-ol | 3391-86-4 | 5.7 | 4.9 | 4.9 | 4.7 | 4.5 | 4.2 | <1 | 8 | 7 | 7 | 7 | 7 | 7 | 6 | mushroom |
| 57 | 2-Furanmethanol | 98-00-0 | <1 | <1 | <1 | <1 | <1 | <1 | <1 | 0 | 0 | 0 | 0 | 1 | 3 | 5 | burnt |
| acids | | | | | | | | | | | | | | | | | |
| 61 | Acetic acid | 64-19-7 | 5.8 | 5.1 | 4.6 | 4.5 | 4.1 | 4 | 2.5 | 10 | 10 | 10 | 10 | 10 | 10 | 10 | sour |
| 63 | Butanoic acid | 107-92-6 | 2.1 | 1.7 | 1.7 | 1.5 | 1.4 | 1.2 | 1.1 | 3 | 3 | 3 | 3 | 3 | 3 | 4 | smelly |
| 67 | Hexanoic acid | 142-62-1 | <1 | <1 | <1 | <1 | <1 | <1 | <1 | 10 | 10 | 10 | 10 | 10 | 10 | 8 | smelly |
| furans | | | | | | | | | | | | | | | | | |
| 82 | 2-Acetylfuran | 1192-62-7 | <1 | <1 | <1 | <1 | <1 | 2.24 | 6.25 | 0 | 0 | 0 | 0 | 1 | 6 | 7 | bitter |

[A] Odor activity value (OAV) ratio of concentration to odor threshold. [B] The number of people who recognize the odor quality through GC-O (total panelists, $n = 10$). [C] Odor quality perceived by GC-O analysis. [D] Not perceived.

Two alcohols (1-octen-3-ol, 2-furanmethanol) and three acids (acetic acid, butanoic acid, hexanoic acid) were identified as aroma-active compounds of jujube, which were found to decrease after microwave treatments, except for 2-furanmethanol. 2-Furanmethanol (burnt and bitter flavor) was the typical product of the Maillard reaction [25], and it was one of the typical flavor substances of dried jujube. With a critical contribution to the flavor of jujube slice samples, the typical C8 alcohol, 1-octen-3-ol, with a mushroom note (also known as "mushroom alcohol"), was detected. Additionally, its concentration decreased with increasing drying time, which may contribute to reducing the floral attribute in jujube slice samples. Acids were identified as the key aroma contributor in jujube samples [36,39]. According to the sensory results, the sour taste was one of the important flavor characteristics of the original jujube slices. Notably, the sour characteristic gradually decreased with the increase in drying time. This change in sensory characteristics was consistent with the trend of acid content. Furthermore, the acid decrease was probably due to the formation of esters and volatilization loss during the drying process [25].

### 3.6. Sensory Analysis

The results of the sensory evaluation analysis of dried jujube slices are shown in the radar charts in Figure 3. In addition to jujube-ID, roasted sweet, caramel, bitter, and burnt were the primary characteristic flavors of jujube slices formed during drying. These flavors together constituted the unique flavor of red jujube after baking. Additionally, the drying time significantly affected the intensities of the basic tastes (roasted sweet, caramel, bitter, and burnt). With the increase in drying time, roasted sweet and caramel increased first and then decreased, and the bitter and burnt flavor increased. For instance, the scores of the B4 sample were the highest in roasted sweet (9.0) and caramel (8.0) attributes. Both 2,3-butanedione and acetoin showed high content and a roasted sweet smell contributing to the "roasted sweet" and "caramel" attributes of the B4 sample. The B7 sample had the highest rated value of the "bitter" and "burnt" descriptors, while the B1 sample showed the lowest sensorial score. This is likely because the adjustment of the treatment drying time resulted in significantly high intensities of roasted sweet and caramel flavors, which was linked to intense drying characteristic aroma. On the other hand, it produced burnt–bitter notes due to excessive heating. At the same time, the range of intensity of the jujube-ID as the important flavor parameter, which represented the flavor of dried jujube, was 9.5–1.0, with a mean value of 4.5; the highest values of this essential attribute were found in dried jujube, and the values then decreased with increasing the time of drying. For instance, samples B6 (30 min) and B7 (45 min) had values of 2.0 and 1.0, respectively. The B1 samples were described more often as having a jujube-ID flavor than the other samples, and acids could perhaps account for the similarity in jujube-ID. Based on the radar charts, the B4 samples were considered caramel and roasted sweet. The overall profiles of the B7 samples were small in the radar charts, given that they were described as bitter and burnt with weak caramel and roasted sweet flavors. The B7 sample showed a strongly bitter note compared to other dried jujube slice samples, mainly because of the high content of 2-acetylfuran (22.40 μg/kg). The different kinds, relative contents, sensory threshold values of aroma-active compounds, and the mutual synergism among aroma-active compounds determined the sensory characteristics of jujube slices. Furthermore, the sensory analysis also agrees with the results of GC-MS and GC-O.

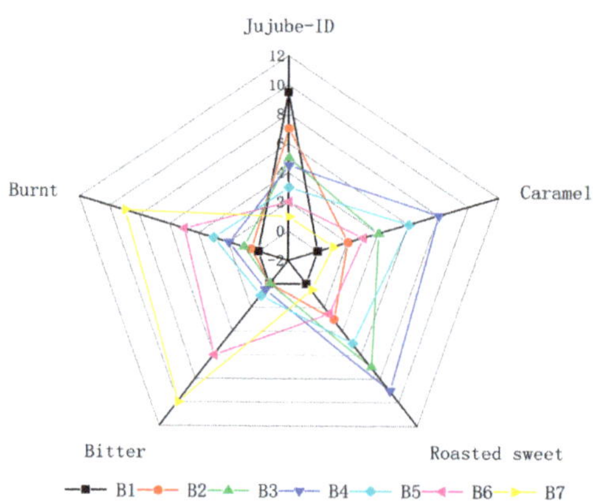


**Figure 3.** Spider chart of sensory evaluation for jujube slices at different microwave drying times: (B1) drying time 0 min; (B2) drying time 1 min; (B3) drying time 2 min; (B4) drying time 2.5 min; (B5) drying time 3 min; (B6) drying time 4 min; (B7) drying time 5 min.

### 3.7. PLSR

To investigate the correlation between the major aroma-active compounds and sensory attributes of dried jujube slices, a PLSR analysis was carried out. The selection of compounds for this analysis is typically based on their relevance and significance in relation to the research objectives. In this case, the 21 aroma-active compounds identified through GC-MS detection and subsequent OAV and DFA analysis were chosen as the predictor variables for the PLSR analysis. These compounds were considered important in terms of their potential contribution to the aroma profile of jujube slices. The 21 aroma-active compounds identified in the jujube slices (Table 2) were subjected to PLSR using PLS1 and PLS2 methods. The X-matrix represented the aroma-active compounds, while the Y-matrix represented one or more sensory attributes. The PLSR2 model extracted two principal components, explaining 97% of cross-validation variance in the X-variables and 92% in the Y-variables. The resulting correlation loading plot is shown in Figure 4, with the inner and outer ellipses indicating 50% and 100% of the explained variance, respectively. All compounds, seven samples, and five sensory attributes were located between the inner and outer ellipses, with r2 values ranging from 0.5 to 1.0, indicating good agreement with the PLSR model. Furthermore, the 21 major aroma-active compounds significantly impacted one or more of the five sensory descriptors.

PLSR1 regression analysis was conducted to identify the odor-active compounds, which significantly contributed to the important sensory attributes of jujube slices, including jujube-ID, caramel, roasted sweet, bitter, and burnt. The results showed that butanoic acid (63), acetic acid (61), hexanoic acid (67), hexanal (2), and nonanal (5) were positively connected with the jujube-ID attribute (Figure 5a). At the same time, 2,3-butanedione (34) and acetoin (35) were positively related to the roasted sweet attribute (Figure 5c).

Moreover, the caramel attribute was significantly positively correlated with γ-butyrolactone (26), 4-cyclopentene-1,3-dione (38), and 4-hydroxy-2,5-dimethyl-3(2H)furanone (45) (Figure 5b). In addition, 2-acetylfuran (82) and 5-methyl-2(5H)-furanone (40) were positively correlated with the bitter attributes (Figure 5d). At the same time, 5-methyl-2-furanaldehyde (10), furfural (7), and 3,5-dihydroxy-2-methyl-4-pyrone (50) also showed a strong correlation with the burnt attribute (Figure 5e). 2-Furanmethanol (57) and 2,3-dihydro-3,5-dihydroxy-6-methyl-4(H)-pyran-4-one (49) had a significant positive impact on the bitter and burnt attribute, which was consistent with findings by Qiao et al. [40]. Moreover, this result was consistent with our previous sensory evaluation, which found that the flavor of caramel and roasted sweet in the samples first increased rapidly and then decreased. At the same time, the bitter and burnt attributes considerably increased, particularly in B6 and B7 samples. The sensory data results also showed good agreement with the twenty-one aroma-active compounds. The sensory differences and similarities in the flavor of samples were mainly caused by the aroma-active compounds' composition, content, and OAV [1].

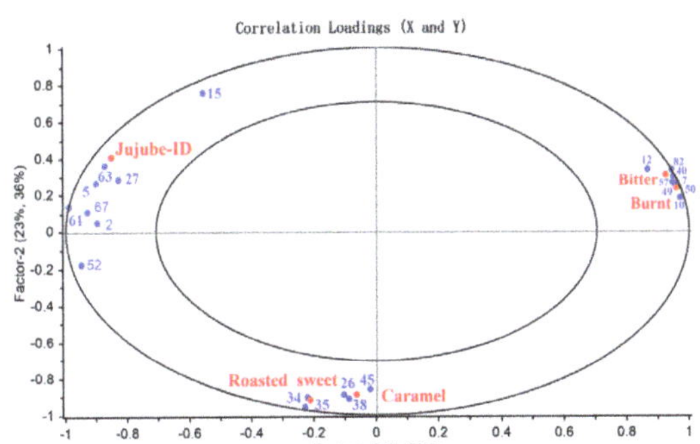


**Figure 4.** Correlation loadings for odor-active compounds with OAVs > 1 (21 compounds in Table 2; the numbering is consistent with the compound numbering in Table 2) and sensory data (Y-matrix) of all samples by PLSR2.

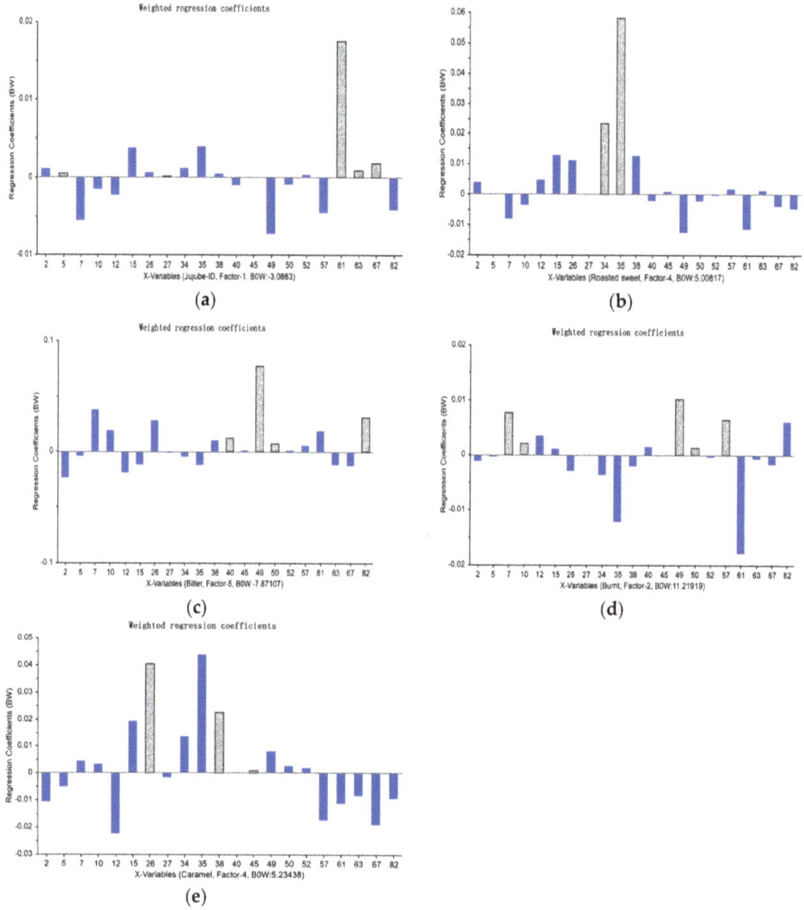


**Figure 5.** Regression coefficients and significance indications for sensory attributes by PLSR1. (**a**) Jujube factor; (**b**) Caramel factor; (**c**) Roasted sweet factor; (**d**) Bitter factor; (**e**) Burnt factor. Gray columns are the key compounds identified by OAV and DFA.

## 4. Conclusions

Through GC-MS detection combined with OAV and DFA analysis methods, we identified the 21 most significant aroma-active compounds in jujube slices, including 17 compounds with OAVs $\geq$ 1, 15 compounds with DFs $\geq$ 5, and 11 compounds with both OAVs $\geq$ 1 and DFs $\geq$ 5, such as caramel-like 4-cyclopentene-1,3-dione and bitter-like 5-methyl-2(5H)-furanone. Simultaneously, some new compounds were found to possess aroma activity in jujube, namely 5-ethyldihydro-2(3H)-furanone,5-methyl-2(5H)-furanone, 3-Hydroxy-2-methyl-4H-pyran-4-one, 2,5-Dimethyl-4-hydroxy-3(2H)-furanone, and 4-Cyclopentene-1,3-dione. Furthermore, the PLSR analysis of 21 aroma-active compounds and sensory evaluation results showed that 2,3-butanedione and acetoin were significantly and positively correlated with the roasted sweet attribute, while γ-butyrolactone, 4-cyclopentene-1, 3-dione, and 2,5-dimethyl-4-hydroxy-3(2H)-furanone provided an intense caramel flavor. Moreover, Ethanone, 1-(2-furanyl)- and 5-methyl-2(5H)-furanone were the primary sources of the bitter taste. Notably, the burnt taste in jujube slices was found to be mainly caused by 3-Hydroxy-2-methyl-4H-pyran-4-one, while 2-Furanmethanol and 2,3-dihydro-3,5-dihydroxy-6-methyl-4H-Pyran-4-one were identified as the primary sources of both burnt and bitter flavors.

Among the 21 flavor compounds, 5-methyl-2(5H)-furanone, 4H-Pyran-4-one, and 2,3-dihydro-3,5-dihydroxy-6-methyl-4H-Pyran-4-one were undetected in the unheated sample (B1). 5-methyl-2(5H)-furanone was also not detected in the sample microwaved for 1 min (B2). Compounds such as 1-(2-furanyl)-ethanone, 5-methyl-2(5H)-furanone, 3-Hydroxy-2-methyl-4H-pyran-4-one, 2-furanmethanol, and 2,3-dihydro-3,5-dihydroxy-6-methyl-4H-pyran-4-one, which exhibit burnt and bitter flavor, gradually increased in content with the increase in microwave time. The contents of compounds (2,3-butanedione, acetoin, $\gamma$-butyrolactone, 4-cyclopentene-1,3-dione, and 2,5-dimethyl-4-hydroxy-3(2H)-furanone) exhibiting roasted sweet and caramel flavors initially increased, then decreased, and reached their highest levels at 2.5 min (B4). Importantly, these conclusions could contribute to understanding the material basis of dried jujube slices' aroma and their change regularities and provide a theoretical basis for improving the aroma quality and regulating peculiar flavor to produce high-quality jujube slices.

**Author Contributions:** Conceptualization, Y.S. and M.T.; methodology, M.T.; software, S.P. and M.T.; validation, X.Z.; formal analysis, X.L.; investigation, X.Y.; resources, M.W.; data curation, S.P. and W.D.; writing—original draft preparation, S.P. and X.Y.; writing—review and editing, X.L.; project administration, M.W.; funding acquisition, Y.S. All authors have read and agreed to the published version of the manuscript.

**Funding:** This research was funded by National Key Research and development project, Taishan Scholar Youth Expert Project, tsqn202211273.

**Data Availability Statement:** The data used to support the findings of this study can be made available by the corresponding author upon request.

**Conflicts of Interest:** The authors declare no conflict of interest.

## References

1. Wang, H.; Li, P.; Sun, S.H.; Zhang, Q.D.; Su, Y.; Zong, Y.L.; Xie, J.P. Comparison of Liquid–Liquid Extraction, Simultaneous Distillation Extraction, Ultrasound-Assisted Solvent Extraction, and Headspace Solid-Phase Microextraction for the Determination of Volatile Compounds in Jujube Extract by Gas Chromatography/Mass Spectrometry. *Anal. Lett.* **2014**, *47*, 654–674. [CrossRef]
2. Available online: https://www.renrendoc.com/paper/238826064.html (accessed on 7 July 2022).
3. Lu, Y.; Bao, T.; Mo, J.; Ni, J.; Chen, W. Research Advances in Bioactive Components and Health Benefits of Jujube (*Ziziphus jujuba* Mill.) Fruit. *J. Zhejiang Univ. Sci. B* **2021**, *22*, 431–449. [CrossRef]
4. Qian, J.; Dai, B.; Wang, B.; Zha, Y.; Song, Q. Traceability in Food Processing: Problems, Methods, and Performance Evaluations-A review. *Crit. Rev. Food Sci. Nutr.* **2022**, *62*, 679–692. [CrossRef]
5. Bao, T.; Karim, N.; Mo, J.; Chen, W. Ultrasound-assisted Ascorbic Acid Solution Pretreated Hot-air Drying Improves Drying Characteristics and Quality of Jujube Slices. *J. Sci. Food Agric.* **2023**, *103*, 4803–4812. [CrossRef] [PubMed]
6. Fathi, F.; Ebrahimi, S.N.; Matos, L.C.; Oliveira, M.B.P.P.; Alves, R.C. Emerging Drying Techniques for Food Safety and Quality: A review. *Compr. Rev. Food Sci. Food Saf.* **2022**, *21*, 1125–1160. [CrossRef]
7. Gao, Q.H.; Wu, C.S.; Wang, M.; Xu, B.N.; Du, L.J. Effect of Drying of Jujubes (*Ziziphus jujuba* Mill.) on the Contents of Sugars, Organic Acids, α-Tocopherol, β-Carotene, and Phenolic Compounds. *J. Agric. Food Chem.* **2012**, *60*, 9642–9648. [CrossRef]
8. Ji, X.; Hou, C.; Yan, Y.; Shi, M.; Liu, Y. Comparison of Structural Characterization and Antioxidant Activity of Polysaccharides from Hujube (*Ziziphus jujuba* Mill.) Fruit. *Int. J. Biol. Macromol.* **2020**, *149*, 1008–1018. [CrossRef]
9. Tan, S.; Wang, Y.; Fu, W.; Luo, Y.; Cheng, S.; Li, W. Drying Kinetics and Physicochemical Properties of Kumquat under Hot Air and Air-impingement Jet Dryings. *Food Sci. Biotechnol.* **2022**, *31*, 711–719. [CrossRef]
10. Bai, R.; Sun, J.; Qiao, X.; Zheng, Z.; Li, M.; Zhang, B. Hot Air Convective Drying of Ginger Slices: Drying Behaviour, Quality Characteristics, Optimisation of Parameters, and Volatile Fingerprints Analysis. *Foods* **2023**, *12*, 1283. [CrossRef] [PubMed]
11. Aprotosoaie, A.C.; Luca, S.V.; Miron, A. Flavor Chemistry of Cocoa and Cocoa Products—An overview. *Compr. Rev. Food Sci. Food Saf.* **2016**, *15*, 73–91. [CrossRef] [PubMed]
12. Chen, Q.; Song, J.; Bi, J.; Meng, X.; Wu, X. Characterization of Volatile Profile from Ten Different Varieties of Chinese Jujubes by HS-SPME/GC–MS Coupled with E-nose. *Food Res. Int.* **2018**, *105*, 605–615. [CrossRef] [PubMed]
13. Song, J.; Chen, Q.; Bi, J.; Meng, X.; Wu, X.; Qiao, Y.; Lyu, Y. GC/MS Coupled with MOS e-nose and Flash GC E-nose for Volatile Characterization of Chinese Jujubes as Affected by Different Drying Methods. *Food Chem.* **2020**, *331*, 127201. [CrossRef] [PubMed]
14. Bi, S.; Xu, X.Y.; Luo, D.S.; Lao, F.; Pang, X.L.; Shen, Q.; Hu, X.S.; Wu, J.H. Characterization of key aroma compounds in raw and roasted peas (*Pisum sativum* L.) by application of instrumental and sensory techniques. *J. Agric. Food Chem.* **2020**, *68*, 2718–2727. [CrossRef] [PubMed]

15. Pu, Y.F.; Ding, T.; Lv, R.L.; Cheng, H.; Liu, D.H. Effect of drying and storage on the volatile compounds of jujube fruit detected by electronic nose and GC-MS. *Food Sci. Technol. Res.* **2018**, *24*, 1039–1047. [CrossRef]
16. Lee, J.E.; Yun, J.H.; Lee, A.R.; Kim, S.S. Volatile components and sensory properties of jujube wine as affected by material preprocessing. *Int. J. Food Prop.* **2018**, *21*, 2052–2061. [CrossRef]
17. Zhang, W.; Dong, P.; Lao, F.; Liu, J.; Liao, X.; Wu, J. Characterization of the major aroma-active compounds in Keitt mango juice: Comparison among fresh, pasteurization and high hydrostatic pressure processing juices. *Food Chem.* **2019**, *289*, 215–222. [CrossRef]
18. Hu, W.; Wang, G.; Lin, S.; Liu, Z.; Wang, P.; Li, J.; Zhang, Q.; He, H. Digital Evaluation of Aroma Intensity and Odor Characteristics of Tea with Different Types-Based on OAV-Splitting Method. *Foods* **2022**, *11*, 2204. [CrossRef]
19. Yang, Y.; Ai, L.; Mu, Z.; Liu, H.; Yan, X.; Ni, L.; Zhang, H.; Xia, Y. Flavor Compounds with High Odor Activity Values (OAV > 1) Dominate the Aroma of Aged Chinese Rice Wine (Huangjiu) by Molecular Association. *Food Chem.* **2022**, *383*, 132370. [CrossRef]
20. Xu, S.; He, W.; Yan, J.; Zhang, R.; Wang, P.; Tian, H.; Zhan, P. Volatomics-assisted Characterization of Aroma and Off-flavor Contributors in Ffresh and Thermally Treated Kiwifruit Juice. *Food Res. Int.* **2023**, *167*, 112656. [CrossRef]
21. Galindo, A.; Noguera-Artiaga, L.; Cruz, Z.N.; Burló, F.; Hernández, F.; Torrecillas, A.; Carbonell-Barrachina, Á.A. Sensory and physico-chemical quality attributes of jujube fruits as affected by crop load. *LWT* **2015**, *63*, 899–905. [CrossRef]
22. Huang, Y.Z.; Liu, Y.; Jin, Z.; Cheng, Q.; Qian, M.; Zhu, B.W.; Dong, X.P. Sensory Evaluation of Fresh/frozen Mackerel Products: A review. *Compr. Rev. Food Sci. Food Saf.* **2021**, *20*, 3504–3530. [CrossRef]
23. Wu, Z.; Zhang, S.; Liu, L.; Wang, L.; Ban, Z. The Grade of Dried Jujube (*Ziziphus jujuba* Mill. cv. Junzao) Affects Its Quality Attributes, Antioxidant Activity, and Volatile Aroma Components. *Foods* **2023**, *12*, 989. [CrossRef] [PubMed]
24. Lu, D.; Zhang, L.; Wu, Y.; Pan, Q.; Zhang, Y.; Liu, P. An Integrated Metabolome and Transcriptome Aapproach Reveals the Fruit Flavor and Regulatory Network during Jujube Fruit Development. *Front. Plant Sci.* **2022**, *13*, 952698. [CrossRef]
25. Zhuang, J.; Xiao, Q.; Feng, T.; Huang, Q.; Ho, C.T.; Song, S. Comparative flavor profile analysis of four different varieties of Boletus mushrooms by instrumental and sensory techniques. *Food Res. Int.* **2020**, *136*, 109485. [CrossRef] [PubMed]
26. Available online: https://www.vcfonline.nl/VcfCompoundSearch.cfm (accessed on 7 September 2022).
27. Available online: http://www.odour.org.uk/index.html (accessed on 7 September 2022).
28. Forero, D.P.; Orrego, C.E.; Peterson, D.G.; Osorio, C. Chemical and sensory comparison of fresh and Dried lulo (*Solanum quitoense* Lam.) fruit aroma. *Food Chem.* **2015**, *169*, 85–91. [CrossRef] [PubMed]
29. Grosshauser, S.; Schieberle, P. Characterization of the Key Odorants in Pan-fried White Mushrooms (*Agaricus bisporus* L.) by Means of Molecular Sensory Science: Comparison with the Raw Mushroom Tissue. *J. Agric. Food Chem.* **2013**, *61*, 3804–3813. [CrossRef]
30. Rychlik, M.; Schieberle, P.; Grosch, W. *Compilation of Odor Thresholds, Odor Qualities and Retention Indices of Key Food Odorants*; Dt. Forschungsanst. für Lebensmittelchemie: Munich, Germany, 1998.
31. Deuscher, Z.; Gourrat, K.; Repoux, M.; Boulanger, R.; Labouré, H.; Le Quéré, J.L. Key Aroma Compounds of Dark Chocolates Differing in Organoleptic Properties: A GC-O Comparative Study. *Molecules* **2020**, *25*, 1809. [CrossRef]
32. Feng, Y.; Cai, Y.; Fu, X.; Zheng, L.; Xiao, Z.; Zhao, M. Comparison of Aroma-active Compounds in Broiler Broth and Native Chicken Broth by Aroma Extract Dilution Analysis (AEDA), Odor Activity Value (OAV) and Omission Experiment. *Food Chem.* **2018**, *265*, 274–280. [CrossRef]
33. Wang, L.N.; Zhu, J.C.; Wang, Y.Q.; Wang, X.; Chen, F.; Wang, X.W. Characterization of Aroma-impact Compounds in Dry. Jujubes (*Ziziphus jujube* Mill.) by Aroma Extract Dilution Analysis (AEDA) and Gas. ChromatographyMass Spectrometer (GC-MS). *Int. J. Food Prop.* **2018**, *21*, 1844–1853. [CrossRef]
34. Hernández, F.; Noguera-Artiaga, L.; Burló, F.; Wojdyło, A.; Carbonell-Barrachina, A.A.; Legua, P. Physico-chemical, nutritional, and volatile composition and sensory profifile of Spanish jujube (*Ziziphus jujuba* Mill.) fruits. *J. Sci. Food Agric.* **2016**, *96*, 2682–2691. [CrossRef]
35. Liu, Y.; Sang, Y.; Guo, J.; Zhang, W.; Zhang, T.; Wang, H.; Cheng, S.; Chen, G. Analysis of Volatility Characteristics of Five Jujube Varieties in Xinjiang Province, China, by HS-SPME-GC/MS and E-nose. *Food Sci. Nutr.* **2021**, *9*, 6617–6626. [CrossRef] [PubMed]
36. Qiao, Y.; Chen, Q.; Bi, J.; Wu, X.; Jin, X.; Gou, M.; Yang, X.; Purcaro, G. Investigation of the Volatile Profile of Red Jujube by Using GC-IMS, Multivariate Data Analysis, and Descriptive Sensory Analysis. *Foods* **2022**, *11*, 421. [CrossRef] [PubMed]
37. Martins, F.C.O.L.; Alcantara, G.M.R.N.; Silva, A.F.S.; Melchert, W.R.; Rocha, F.R.P. The role of 5-hydroxymethylfurfural in food and recent advances in analytical methods. *Food Chem.* **2022**, *395*, 133539. [CrossRef] [PubMed]
38. Zhu, J.; Xiao, Z. Characterization of the Major Odor-Active Compounds in Dry Jujube Cultivars by Application of Gas Chromatography–Olfactometry and Odor Activity Value. *J. Agric. Food Chem.* **2018**, *66*, 7722–7734. [CrossRef]
39. Egea, M.B.; Bertolo, M.R.V.; Oliveira Filho, J.G.; Lemes, A.C. A Narrative Review of the Current Knowledge on Fruit. Active Aroma Using. Gas. Chromatography-Olfactometry (GC-O) Analysis. *Molecules* **2021**, *26*, 5181. [CrossRef]
40. Qiao, Y.N.; Bi, J.F.; Chen, Q.Q.; Wu, X.Y.; Jin, X.W.; Gou, M. Rapid and Sensitive Quantitation of DDMP (2,3-dihydro-3,5-dihydroxy-6-methyl-4H-pyran-4-one) in Baked Red Jujubes by HS-SPME-GC-MS/MS. *Food Control* **2022**, *135*, 108820. [CrossRef]

foods

MDPI

*Article*

# Effect of Heat Treatment on the Digestive Characteristics of Different Soybean Oil Body Emulsions

**Xufeng Yang [1,2], Luyao Zhou [1,2], Yingying Wu [1,2], Xiuzhen Ding [1,2], Wentao Wang [1,2], Dajian Zhang [3,*] and Luping Zhao [1,2,*]**

1 College of Food Science and Engineering, Shandong Agricultural University, Tai'an 271018, China; 17863852660@163.com (X.Y.); 15194109917@163.com (L.Z.); yying1220@126.com (Y.W.); xzd@sdau.edu.cn (X.D.); wangwt@sdau.edu.cn (W.W.)

2 Engineering and Technology Center for Grain Processing of Shandong Province, Tai'an 271018, China

3 State Key Laboratory of Crop Biology, Shandong Agricultural University, Tai'an 271018, China

* Correspondence: dajianzhang@sdau.edu.cn (D.Z.); zhaolp0126@126.com (L.Z.); Tel.: +86-0538-8246009 (L.Z.)

**Abstract:** Soybean oil body (SOB) emulsions were prepared using OBs extracted at pH 11.0 and pH 7.0. The pH 11.0-SOB comprised oleosins, whereas pH 7.0-SOB comprised extrinsic proteins and oleosins. All SOB emulsions were heated at 60–100 °C for 15 min. Heating may lead to the release of extrinsic proteins from the surface of pH 7.0-SOB due to heat-induced denaturation. The total proportion of α-helix and β-sheets gradually decreased from 77 (unheated) to 36.2% (100 °C). During stomach digestion, the extrinsic protein hydrolysis of heated pH 7.0-SOB emulsions was fast between 60 and 80 °C, and it then slowed between 90 and 100 °C; heating inhibited the oleosin hydrolysis of pH 7.0- and 11.0-SOBs. Heat treatment promoted aggregation and coalescence, and it resulted in increased particle sizes for all emulsions. Larger aggregates were found in heated pH 7.0-SOB emulsions, and larger oil droplets were found in heated pH 11.0-SOB emulsions. After intestinal digestion, the droplets of all SOB emulsions gradually dispersed, and particle sizes decreased. Different heating temperatures had lesser effects on particle sizes and microstructures. Lipolysis was affected by the extraction pH and heating. For pH 11.0-SOB emulsions, the FFA release tendency was greatly affected by the heating temperature, and heating to 80 °C resulted in the highest FFA release (74%). However, all pH 7.0-SOB emulsions had similar total FFA releases. In addition, the droplet charges of heated pH 7.0-SOB emulsions were lower than those of unheated pH 7.0-SOB emulsions in both the intestine and stomach phases; however, the charge changes in different pH 11.0-SOB emulsions showed the opposite tendency. This study will offer guidance regarding the application of SOB emulsions in food.

**Keywords:** heat treatment; soybean oil body; protein hydrolysis; digestive characteristics; FFA release

## 1. Introduction

Soybeans are widely cultivated as oil crops across the world. The protein and lipid contents of soybeans are 37.7–44.5% and 17.2–21.6%, respectively [1]. It is known that protein storage vacuoles act as storage organelles for soybean proteins, and soybean oil is stored in special organelles known as oil bodies (SOBs). A SOB is composed of a core of triglycerides (TAGs) coated with a monolayer of phospholipids that is embedded into the OB's intrinsic proteins [2–4]. Intrinsic proteins include steroleosins, caleosins, and oleosins, and they play significant roles in resisting environmental pressure and maintaining the SOBs' integrity [5]. Oleosins are small molecules with molecular weights of 15–26 kDa (mainly 24 kDa oleosins, 18 kDa oleosins, and 16 kDa oleosins). The extrinsic proteins include glycinin (11S: acid polypeptides, A; basic polypeptides, B), β-conglycinin (7S: α′, α, β), γ-conglycinin, Bd 30K, and P34 [6]. SOBs also contain phytosterol [7] and tocopherol [8], which are components required for healthy functioning. SOBs have great oxidation stability and dispersion stability due to their functional components and special structures. SOB emulsion preparation does not need the involvement of other emulsifiers or homogenization. It also conforms to food processing requirements related to safety and environmentally friendly practices.

Due to the need for food processing and storage, heat treatment is one of the most common techniques used in the field of food. Furthermore, heating is a common method used to increase the digestibility of proteins in various foods [9]. However, some studies found that heating could change the interfacial protein compositions and structures of OBs. Yan et al. (2016) found that extrinsic proteins/oleosins in soymilk were approximately 1.1 in unheated conditions [10]. After heating, the extrinsic proteins/oleosins were 0.72 (70 °C, 8 min), 0.43 (80 °C, 4 min), 0.34 (90 °C, 4 min), and 0.31 (100 °C, 4 min), respectively, indicating that the extrinsic proteins remained within the heated SOBs. As the SOBs were extracted at 60–100 °C for 30 min, the extrinsic proteins decreased from 65.29 to 4.74% as the heating temperatures increased [11]. In addition, the oleosin of the raw camellia OB was 13.3 kDa. The bands of 25–30 kDa and 55 kDa were determined based on the extended roasting time. The protein bands of the boiled OBs became shallow [12]. The study found that the secondary structure of the raw camellia OB interface proteins changed [12]. The random coil increased from 6% (raw camellia OB) to 13–18% (roasted seeds treated at 130 °C, for 5–10 min) or 13–23% (seeds boiled for 20–60 min). However, the β-sheet decreased from 25% (raw camellia OB) to 20% (roasting) or 5% (boiling). Fu et al. (2020) studied the effects of pre-heating treatments (including heating raw soymilk at 65 °C, 75 °C, and 85 °C for 30 min, as well as ultra-high temperature-treated dry soybeans at 110 °C, 120 °C, and 130 °C for 10 s) on SOBs [13]. They found that heating soymilk led to storage protein denaturing and partial unfolding, whereas ultra-high temperatures could lead to large amounts of soluble aggregates being present within oleosins and storage proteins emerging via disulfide bonds. Therefore, the interfacial protein structures and compositions of OBs were affected by heat treatment.

OBs could be considered as special protein-stabilized oil droplets because the entire surface was covered by proteins [14]. We considered the properties of the OBs' surfaces to be similar to those of protein-stabilized emulsions. Some studies reported that heat treatment could affect the digestive characteristics of protein-stabilized emulsions and OB emulsions. Capuano et al. (2018) investigated the behavior of raw and roasted hazelnut OBs via in vitro lipid digestion [15]. They found that oleosins of raw hazelnut OBs were hydrolyzed to produce an 8.0-dalton pepsin-resistant peptide after gastric digestion, and the free fatty acid (FFA) releases of raw and roasted hazelnut OBs were 31% and 45%, respectively. Ding et al. (2020) studied the digestive characteristics of SOBs that were extracted and heated (95 °C, 15 min), along with non-heated raw soybeans, via in vitro gastrointestinal digestion [16]. It is worth noting that the aggregation statuses of droplets for both heated and non-heated SOBs are similar in the stomach phase. Conversely, the FFA release rate of heated SOBs was higher (23.01%) than that of non-heated SOBs (21.05%). Ye et al. (2020) studied the digestive behaviors of whey protein-stabilized emulsions treated using heat treatment, and they found that the heated whey protein-stabilized emulsion had a higher degree of protein hydrolysis [17]. During in vitro intestinal digestion, the FFA releases of the heated emulsion and unheated emulsion were 65% and 52%, respectively. Therefore, we postulated that heat treatment changed the structures and compositions of OB interfacial proteins, which could affect the digestive regularity of OB emulsions. However, the digestive characteristics of SOB emulsions treated at different heating temperatures were not systematically investigated. In addition, we previously reported that the digestive characteristics of SOBs were different [18]. Therefore, the digestive characteristics of pH 7.0- and 11.0-SOBs that were heat treated needed to be studied.

In this study, pH 7.0- and pH 11.0-SOB emulsions were each treated at 60 °C, 70 °C, 80 °C, 90 °C, and 100 °C. The digestive regularities of pH 7.0- and pH 11.0-SOB emulsions were researched in the gastrointestinal stage, including protein hydrolysis, zeta potentials, particle sizes, microstructures, and FFA releases. This study will offer theoretical guidance for the application of SOB emulsions in food, such as the delivery of functional nutrients and dietary lipid.

## 2. Materials and methods

### 2.1. Materials

We bought soybean seeds (Zhonghuang 37) from the local market (Tai'an, China). Protein markers (14.4–97.4 kDa) were bought from Solarbio (Beijing Solarbio Science and Technology Co., Ltd., Beijing, China). Pancreatin, pepsin, Fast Green and Nile Red, and bile salts were bought from Sigma-Aldrich (St. Louis, MO, USA). All chemical reagents were analytical grade.

### 2.2. Extractions of pH 11.0-SOB and pH 7.0-SOB

The SOBs were extracted using the modified methods described by Han et al. (2020) [19]. Soybean seeds (50 g) were washed with deionized (ID) water to remove dust and soaked in fresh ID water (300 g) for 16 h at 4 °C in a refrigerator. The swollen seeds were collected and washed three times with fresh ID water. The pre-cooled (4 °C) ID water was mixed with the swollen seeds (ID water/seeds = 9/1, *w/w*), and the mixture was ground via a crushing machine (BLST4090-073, Qster) at 22,000 r/min for 2 min. Then, four layers of gauze were used to filter the resulting slurry. Sucrose (62.5 g) was added to the filtrate (312.5 g) and mixed well. The mixture was divided into two parts, and their pHs were adjusted to 11.0 and 7.0, respectively, using 0.2–2 mol/L KOH and HCl solutions. The mixtures were centrifuged at 43,667× *g* and 4 °C for 30 min. The floats were collected from the upper layer and mixed again with fresh ID water. The pHs of the mixtures were adjusted to 11.0 and 7.0, respectively, and they were centrifuged under the same conditions. The above steps were repeated twice. The obtained SOBs were called pH 11.0-SOB and pH 7.0-SOB, respectively.

### 2.3. Emulsion Preparation and Heat Treatment

SOBs (8.0 g) were dispersed into the ID water to prepare the SOB emulsions (5%, dry basis). All emulsions were sheared (3000 r/min, 2 min) using a dispersion machine (T25 Digital Package 2, German IKA Company, Königswinter, Germany). The pH 7.0- and pH 11.0-SOB emulsions were heated at 60 °C, 70 °C, 80 °C, 90 °C, and 100 °C for 15 min and immediately cooled. An unheated SOB (4 °C) was used as the control. Some samples were centrifuged (11,102× *g*) again, and the floats were analyzed using Tricine-SDS-PAGE. $NaN_3$ (0.2%, *w/w*) was added into the other SOB emulsions, which were immediately used for further analysis.

### 2.4. In Vitro Gastrointestinal Digestion of SOBs

The digestive model was designed according to the method previously mentioned by Yang et al. (2022) [18]. Each SOB emulsion (80.0 g) was mixed with 20 mL of simulated gastric fluid SGF (the SGF contained 0.43 mol/L of NaCl). The pH of each mixture was controlled to pH 4.0 using HCl (0.2–2 mol/L), and pepsin (0.68 mg/mL) was added to each mixture. The simulated stomach digestion occurred under the conditions of pH 4.0 and 37 °C for 2 h.

When simulated stomach digestion was completed, the pH of each mixture was adjusted to 7.0 with KOH (0.2–2 mol/L). The simulated intestinal fluid (SIF) was composed of CaCl2 (0.3 mol/L), KCl (25 mmol/L), and NaCl (0.75 mol/L). Bile salts (0.62 g) were added into the prepared SIF (57.6 mL). Then, pancreatin was added into the solution, and the final concentration was 6.35 mg/mL. This solution simulated intestinal digestion at pH 7.0 and 37 °C, which lasted for 3 h.

### 2.5. Zeta Potential

Samples were collected at 120 min (termed G120) in the stomach phase, cooled, and diluted 400 times using citric acid solution (20 mmol/L, pH 4.0). In addition, samples were collected at 180 min (termed I180) in the intestine phase and cooled. The initial emulsions and all samples were also diluted for the number of times described above using phosphate solution (20 mmol/L, pH 7.0). Zeta potentials were carried out using a Laser Nano-ZS Particle Analyzer (Malvern Instruments Ltd., London, UK) at 25 °C [19,20].

### 2.6. Droplet Size Analysis

Samples from stomach and intestinal digestion were diluted 1000 times using a citric acid solution (20 mmol/L, pH 4.0) and a phosphate solution (20 mmol/L, pH 7.0), respectively. The particle sizes were determined using a Laser Nano-ZS Particle Analyzer (Malvern Instruments Ltd., London, UK) at 25 °C.

### 2.7. Tricine-SDS-PAGE

Tricine-SDS-PAGE was conducted according to the method of Zhao et al. (2016) [21], albeit with modifications. The concentrations of stacking and separating gels were 4% and 16%, respectively. The protein concentrations of all samples were adjusted to 1.5 mg/mL via mixing with Tricine-SDS-PAGE sample buffer. β-mercaptoethanol (2%, *v/v*) was added to all samples, and the mixed solutions were heated at 100 °C for 5 min. Then, 8 μL of each sample was loaded into the sample well, and samples were electrophoresed at a constant voltage of 30–35 mV until all samples entered into the stacking gel, and the constant voltage was then changed to 100 mV until complete. The gel was stained using Coomassie Brilliant Blue G-250. The intensities of different protein bands were analyzed via Image Lab 3.0 Software (Bio-Rad, Hercules, CA, USA).

### 2.8. Microscope Observation

Microstructures were observed using a Confocal Laser Scanning Microscope (CLSM) 880 (Carl Zeiss, Jena, Germany). The oils and proteins were stained using 0.02% (*w/v*) Nile Red dissolved in 10% (*v/v*) acetone and 0.1% (*w/v*) Fast Green aqueous solution, respectively. Samples (200 μL) were stained with Nile Red (20 μL) and Fast Green (25 μL) and stored with restricted lighting for 10 min [22,23]. The excitation wavelengths of Nile Red and Fast Green were 488 nm and 633 nm, respectively, which were detected at 580–620 nm and 651–710 nm. Samples were observed with 43× oil lenses. Zeiss Zen 2012 Software (Jena, Germany) was used to analyze all images.

### 2.9. Free Fatty Acid (FFA) Release

During intestinal digestion, pH-STAT automatic potentiometric titration apparatus was used to monitor the consumed amount of KOH (0.1 mol/L) [24,25]. FFA release was calculated on the basis of KOH consumption, as per the below equation:

$$FFA\ (\%) = \frac{V_{KOH} \times m_{KOH} \times M}{M_{lipid} \times 2} \times 100 \quad (1)$$

$V_{KOH}$ and $m_{KOH}$ represent the consumed volume and content of KOH, respectively; $M$ represents the molecular weight of SOBs, i.e., 876.56 g/moL; and $M_{lipid}$ represents the weight of TAGs.

### 2.10. Statistical Analysis

Experiments were performed on 3 separate occasions in triplicate using freshly prepared samples. The data were presented as the means with standard deviations. Statistical analysis was performed using SPSS statistics version 25 following an ANOVA model (IBM Inc., Armonk, NY, USA). A confidence level of $p < 0.05$ was considered to be statistically significant.

## 3. Results and Discussion

### 3.1. Protein Compositions and Structures of pH 7.0- and pH 11.0-SOB Emulsions

The protein compositions of unheated and heated pH 7.0- and pH 11.0-SOB emulsions are shown in Figure 1a,c. The unheated pH 7.0-SOB had oleosins (24 kDa, 18 kDa, and 16 kDa) and extrinsic proteins (main 7S: α, α′, and β; 11S: A, A3 and B). However, the unheated pH 11.0-SOB only had oleosins. Compared to unheated SOBs, the protein compositions of the pH 7.0- and pH 11.0-SOB emulsions heated at 60 °C–100 °C showed no significant changes. In order to determine the proteins bound to SOBs, the heated pH 7.0- and pH 11.0-SOB emulsions were centrifuged, and their protein compositions are shown in Figure 1b,d. It was found that all 7S and 11S were released from the pH 7.0-SOB heated to more than 60 °C; however, oleosins were still bound to the heated pH 7.0- and 11.0-SOB. Oleosins formed "T" shapes on the SOB surfaces. The N-terminal and C-terminal regions of the peptide chain were exposed to the outside, and the hydrophobic region in the middle of the peptide chain was embedded firmly inside of the SOB [5]. Therefore, extrinsic proteins were released at ≥60 °C.

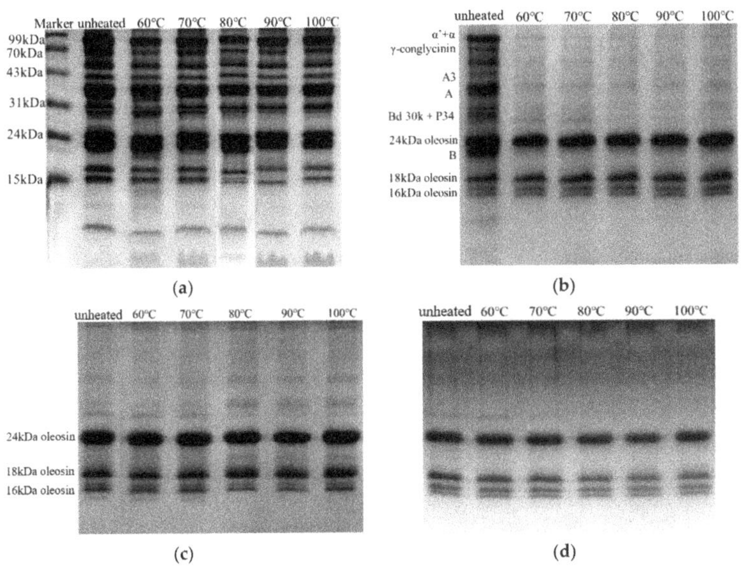


**Figure 1.** Tricine–SDS–PAGE protein profiles of pH 7.0-SOB (**a**,**b**) and 11.0-SOB (**c**,**d**) treated using different heat treatments. (**a**,**c**) represents heated SOBs; (**b**,**d**) represents the floats obtained by centrifuging heated SOB emulsions.

The Raman spectra of the pH 7.0-SOB and the secondary structures of the interfacial proteins are shown in Figure 2. The peaks featured in the analysis of the protein structure were referred to as amide I (1600–1700 $cm^{-1}$) and amide III (1230–1300 $cm^{-1}$). In order to reveal the protein structure of the pH 7.0-SOB, we quantitatively fitted amides I and III via the Raman spectrum (Figure 2b). The secondary protein structure of the unheated pH 7.0-SOB was mainly made up of 45.6% α-helix and 31.4% β-sheet. However, the α-helix gradually decreased from 45.6 to 19.5% with increases in the heating temperatures, and the random coil increased from 18 (60 °C) to 37.5% (100 °C). In addition, the total proportion of α-helix and β-sheet decreased from 72 (unheated) to 36.2% (100 °C). This result indicated that heating could gradually change the protein's structure from an ordered structure to a disordered structure, which explained the release of 7S and 11S from the pH 7.0-SOB surface (Figure 1b). Some studies reported that α′/α and β were separated from 7S, and A and B were separated from 11S by heating raw soybean milk, resulting in 7S and 11S being released from the SOBs [26,27].

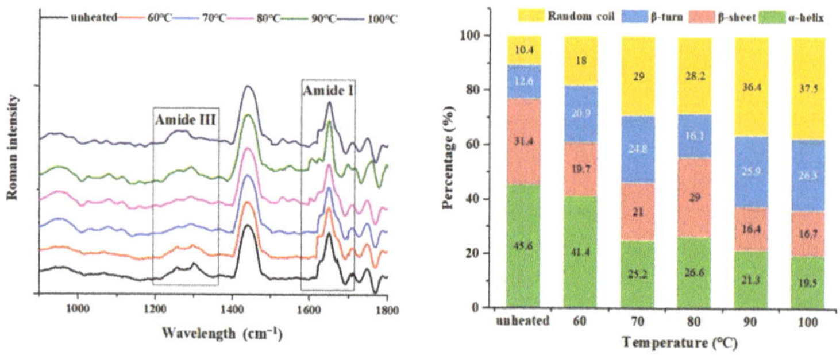


(**a**) (**b**)

**Figure 2.** Confocal Raman spectra of SOB interface proteins (**a**), and the secondary structure determined via curve-fitting of Raman spectra (**b**).

## 3.2. Protein Hydrolysis in Stomach Digestion

Protein hydrolysis was investigated in the stomach digestion phase, and the results are shown in Figures 2 and 3. All protein bands of each SOB emulsion became gradually shallower and produced low molecular bands with the extension of the digestive time, which indicated that proteins were gradually hydrolyzed and produced small molecular peptides.

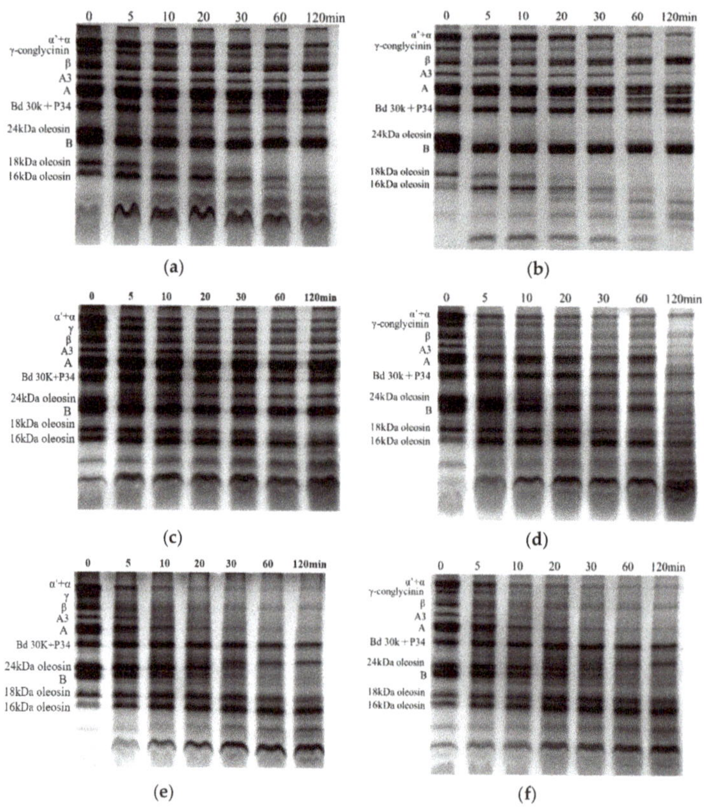


**Figure 3.** Tricine-SDS-PAGE of pH 7.0-SOB during the stomach phase. (**a**–**f**), unheated, 60 °C, 70 °C, 80 °C, 90 °C, 100 °C.

The effects of heat treatment on extrinsic protein and oleosin hydrolysis were greatly different for pH 7.0-SOB emulsions. The extrinsic proteins of the heated pH 7.0-SOB emulsions were more rapidly hydrolyzed than those of the unheated pH 7.0-SOB after 120 min (Figure 3), indicating that the released extrinsic proteins were more easily hydrolyzed than the extrinsic proteins bound to the unheated pH 7.0-SOB. The hydrolysis of extrinsic proteins became faster from 60 to 80 °C, before slowing from 90 to 100 °C. Fu et al. (2020) reported that the 7S and 11S denaturation temperatures were about 70 °C and 90 °C, respectively [13]. Under their denaturation temperatures, 7S and 11S were partially denatured, which probably exposed the active groups and enzyme binding sites inside the molecule. However, 7S and 11S were completely denatured at 90 °C–100 °C, resulting in the thermal aggregation of extrinsic proteins, and the exposed enzyme binding sites were covered again [17]. Compared to the unheated pH 7.0-SOB emulsion, heat treatment inhibited oleosin hydrolysis. For example, 24 kDa of oleosin in the unheated pH 7.0-SOB was completely hydrolyzed at 5 min, and hydrolysis of 18 kDa of oleosin required 30 min; for 100 °C, 24 kDa of oleosin was completely hydrolyzed at 30 min, while 18 kDa of oleosin was still not completely hydrolyzed at 120 min. In addition, oleosin hydrolysis rates showed a tendency to decrease with the increases in the heating temperatures for pH 7.0-SOB emulsions. Interestingly, the amount of 16 kDa first showed an increasing tendency and then decreased, as 24 kDa of oleosin was hydrolyzed and produced 16 kDa of peptide [4,28]. Therefore, heating greatly affected the protein hydrolysis of pH 7.0-SOB emulsions.

Heat treatment also affected oleosin hydrolysis for pH 11.0-SOB emulsions. For the unheated pH 11.0-SOB emulsion, 24 kDa of oleosin was completely hydrolyzed at 5 min, and 18 kDa of oleosin was hydrolyzed at 20 min. At 100 °C, most of the 24 kDa of oleosin was hydrolyzed at 60 min, and most of the 18 kDa of oleosin was hydrolyzed at 120 min. It was suggested that heating inhibited oleosin hydrolysis. In addition, the oleosin hydrolysis rate decreased with the increases in the heating temperatures, which was similar to the result for the pH 7.0-SOB (Figure 4). Small pepsin-resistant peptides (marked in black) were produced from oleosins in the stomach phase in all pH 11.0-SOB emulsions.

### 3.3. Zeta Potential

It was shown that the initial pH 7.0- and pH 11.0-SOB emulsions had negative charges, and their zeta potentials were positive in the stomach phase and changed to negative during intestinal digestion (Figure 5), which occurred due to pH values of 4.0 and 7.0 in the gastric and intestinal environments, respectively. Before digestion, the zeta potentials of the unheated pH 7.0- and 11.0-SOB were −21.7 mV and −15.67 mV, respectively, which was due to the extrinsic proteins bound to the pH 7.0-SOB. This result was consistent with Yang et al. (2022), and they reported that the zeta potential of SOB emulsions increased with the increases in the extraction pHs from 5.0 to 11.0 [18]. In addition, the zeta potentials of pH 7.0-SOB emulsions decreased from −18.9 mV at 60 °C to −22.83 mV at 70 °C, before increasing to −11.15 mV at 100 °C (Figure 5a), which might be relative to structures of the interfacial proteins. Heating temperatures (60 °C–100 °C) had no effects on the zeta potentials of pH 11.0-SOB emulsions because heated pH 11.0-SOB emulsions had the same protein compositions (Figure 1c,d).

After stomach digestion for 120 min, the zeta potentials of heated pH 7.0-SOB emulsions decreased significantly compared to the unheated pH 7.0-SOB (Figure 5a). This discrepancy was attributed to the faster hydrolysis of extrinsic proteins in heated pH 7.0-SOB emulsions (Figure 3). Zeta potentials decreased with the increase in the heating temperature from 60 to 80 °C, before increasing with the increase from 80 to 100 °C. These changes were consistent with the regularity of extrinsic protein hydrolysis in different heated pH 7.0-SOB emulsions (Figure 3). It was revealed that extrinsic protein hydrolysis had important effects on the zeta potentials of pH 7.0-SOB emulsions. In addition, the zeta potentials of unheated pH 11.0-SOB emulsions were lower than those of heated pH 11.0-SOB emulsions after stomach digestion. This result was attributed to the fact that more oleosins were hydrolyzed for unheated pH 11.0-SOB emulsions. The zeta potentials of pH 11.0-SOB emulsions experienced no significant changes with the increase in temperature from 60 to 100 °C. After intestinal digestion for 180 min, the zeta potentials of all heated pH 7.0-SOB emulsions were slightly higher than those of unheated pH 7.0-SOB emulsions, and the zeta potentials were highest when heated at 80 °C. However, different heated pH 11.0-SOB emulsions had similar zeta potentials, which were significantly lower than those of unheated pH 11.0-SOB emulsions. McClements and Li. (2010) reported that small molecular substances (such as FFAs, bile salts, and peptides) were absorbed via the droplet interface [29], which increased droplet charges. It is possible that different amounts of small molecular substances were absorbed via the droplet interface.

As a result, the zeta potential of the unheated pH 11.0-SOB emulsion was higher than that of the pH 7.0-SOB with extrinsic proteins before digestion, and the heating temperature had a greater effect on the zeta potential of the pH 7.0-SOB. Moreover, the heated pH 7.0-SOB emulsions had fewer droplet charges than those of the unheated pH 7.0-SOB in both the intestine and stomach phases; however, the charge changes in pH 11.0-SOB emulsions showed the opposite tendency. The heating temperatures (60–100 °C) had no obvious influence on the zeta potentials of pH 11.0-SOB emulsions.

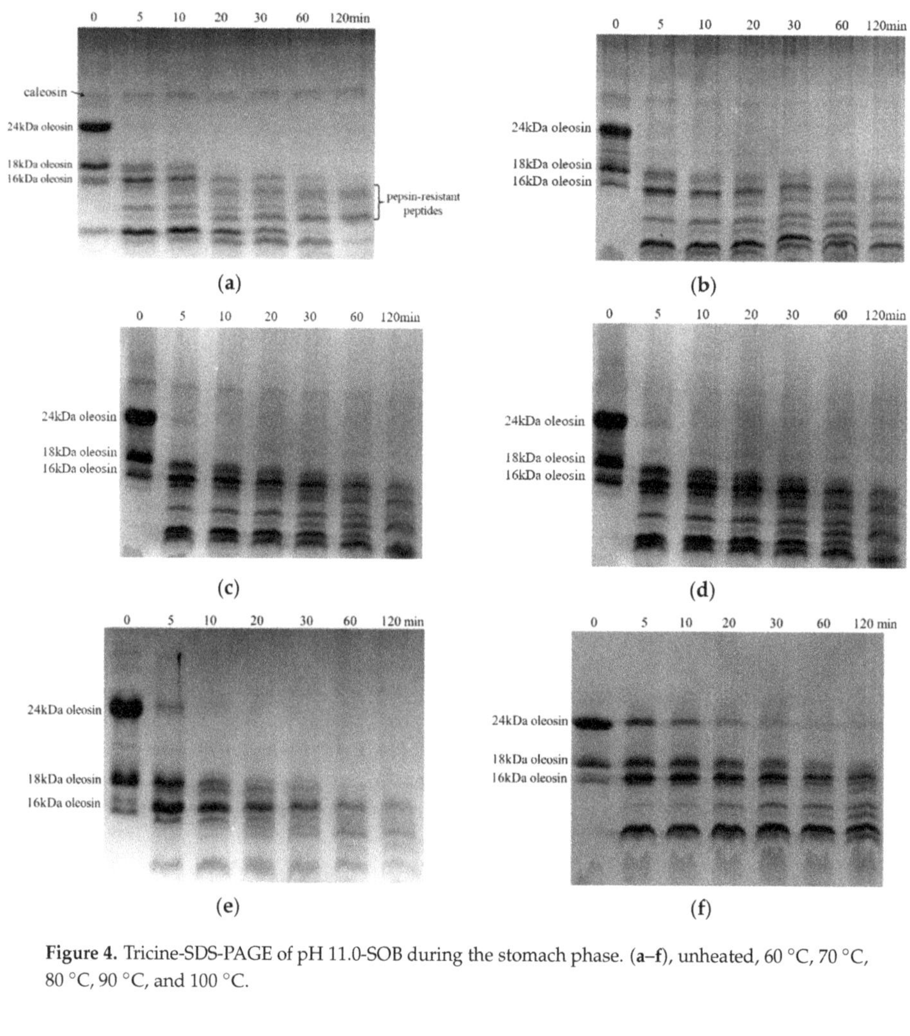


**Figure 4.** Tricine-SDS-PAGE of pH 11.0-SOB during the stomach phase. (**a**–**f**), unheated, 60 °C, 70 °C, 80 °C, 90 °C, and 100 °C.

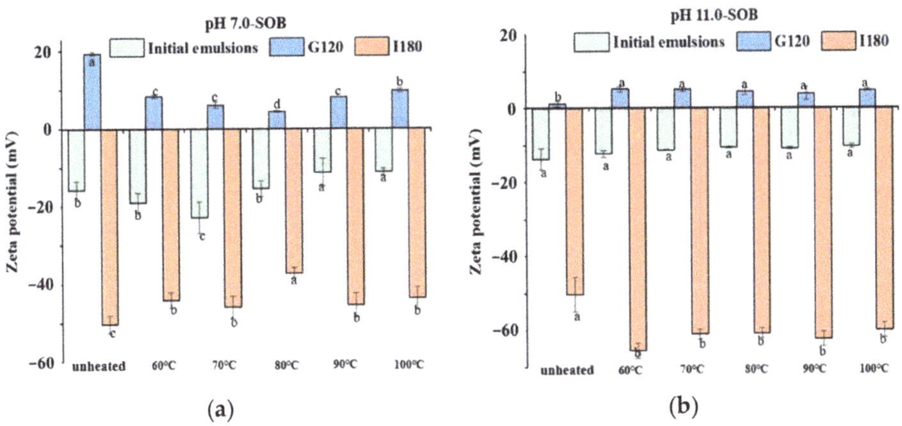


**Figure 5.** Zeta potential of pH 7.0-SOB (**a**) and 11.0-SOB (**b**) emulsions treated using different heat treatments during gastrointestinal digestion. Different small letters represent significant differences ($p < 0.05$) in the same digestion stage. G120 represents SOB emulsions incubated for 120 min during the stomach phase. I180 represents SOB emulsions incubated for 180 min during the intestinal phase. The following figure is similar.

### 3.4. Particle Size

Before digestion, the average particle size of the unheated pH 7.0-SOB emulsion was 658 nm, and the particle sizes of heated pH 7.0-SOB emulsions first decreased to 459 nm (60 °C), before remaining constant at about 327 nm (70–100 °C). This result occurred due to extrinsic proteins released from the pH 7.0-SOB surface via heating (Figure 1a,b). Conversely, there were no significant changes in the particle sizes of unheated and heated pH 11.0-SOB emulsions because the unheated and heated pH 11.0-SOB emulsions had the same protein composition (Figure 1c,d).

It was shown that the particle sizes of pH 7.0- and pH 11.0-SOB emulsions increased significantly in the stomach phase (Figure 6). The reason for this increase was that the pH of gastric fluid was close to the isoelectric points of SOB interfacial proteins, which further reduced electrostatic repulsion and droplet aggregation [30]. Moreover, a coalescence of droplets occurred due to the hydrolysis of extrinsic proteins and oleosins [31]. Compared to the unheated pH 7.0-SOB, the particle sizes of all heated pH 7.0-SOB emulsions were larger, and this result might be attributed to the fact that more aggregates formed due to weak electrostatic repulsion (Figure 5a). However, the particle sizes of heated pH 11.0-SOB emulsions with more positive charges were higher than those of unheated pH 11.0-SOB emulsions with fewer positive charges. It was possible that the droplet coalescence had greater effects on particle size than on aggregation. When heating temperature increased from 60 to 100 °C, the droplet sizes of pH 11.0-SOB emulsions showed a tendency to decrease. This result might be attributed to the fact that oleosin hydrolysis became slower, and the degree of droplet coalescence decreased with the increase in the heating temperature. Compared to stomach digestion, the particle sizes of all pH 7.0-and pH 11.0-SOB emulsions significantly decreased due to droplets having more negative charges after intestinal digestion for 180 min. There were no regular changes in particle sizes for pH 7.0-SOB emulsions in the intestinal phase. Compared to the particle sizes in unheated emulsions, heating at 60 °C and 70 °C, and 80–100 °C decreased the particle sizes of pH 11.0-SOB emulsions. Therefore, the heating temperatures had more effect on particle sizes in the stomach phase than in the intestinal phase.

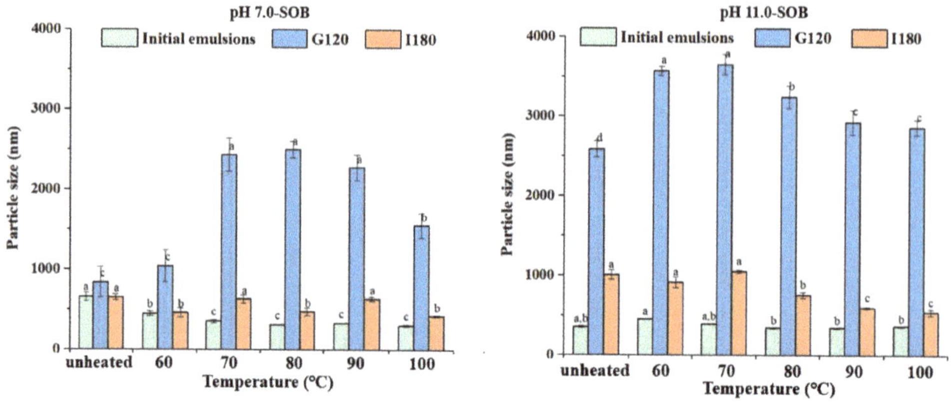


(a) (b)

**Figure 6.** Particle size of pH 7.0-SOB (**a**) and 11.0-SOB (**b**) emulsions treated at different treatments during gastrointestinal digestion. Different small letters represent significant differences ($p < 0.05$) in the same digestion stage.

### 3.5. Microstructures

The droplets were observed using CLSM, and oil phases and proteins were marked with red and green fluorescent dyes (Figure 7). Before digestion, the droplets of all initial pH 11.0- and 7.0-SOB emulsions dispersed well. During the stomach phase, the aggregation degree of droplets in pH 7.0-SOB emulsions first increased, before decreasing with the increases in the heating temperature, and more and larger aggregates could be observed in pH 7.0-SOB emulsions heated at 80 °C. This result was consistent with the particle size analysis (Figure 6a). In addition, small oil droplets were present in pH 7.0-SOB emulsions, suggesting that slight droplet coalescence occurred due to oleosin and extrinsic protein hydrolysis. Aggregates and large oil droplets occurred in heated pH 11.0-SOB emulsions during stomach digestion. These results revealed that aggregation and coalescence were the reasons for the increase in particle sizes of all emulsions in the stomach phase (Figure 6), and heat treatment resulted in larger oil droplets in pH 11.0-SOB emulsions and larger aggregates in pH 7.0-SOB emulsions. These aggregations and the coalescence induced by heating might increase steric hindrance, and oleosin hydrolysis was inhibited (Figures 3 and 4). After intestinal digestion, the droplets gradually dispersed, the particle sizes of all SOB emulsions showed decreasing tendencies, and microstructures showed no obvious changes in the different SOB emulsions. Thus, the heating temperatures and extraction pHs of SOB emulsions had more influence on microstructures in the stomach phase than those in intestinal digestion.

### 3.6. The Release of FFAs

During intestinal digestion, TAGs were hydrolyzed using pancreatin, and the total FFA release increased gradually with increasing digestive time (Figure 8). The FFA release rate of unheated and heated pH 7.0-SOB emulsions first experienced a rapid increase, followed by a flat curve. For the pH 7.0-SOB emulsion, the FFA release rate increased with increases in the heating temperatures for 0–10 min, and the total FFA release was similar (about 70%) at 180 min. In addition, FFA release was greatly affected by the heating temperatures in the pH 11.0-SOB emulsion. For unheated and heated (60 °C and 70 °C) pH 11.0-SOB emulsions, the FFA release curve showed a transient lag phase during initial digestion, and the FFA content increased with prolonged digestive time at similar rates. The FFA release of the pH 11.0-SOB emulsion was heated at 80 °C, 90 °C, and 100 °C; it first rapidly increased, before continuing to increase more slowly. Finally, the total FFA releases were 60%, 65.25%, 64.5%, 74%, 62.1%, and 70.3% for pH 11.0-SOB emulsions, which were unheated and heated at temperatures in the range 60–100 °C, respectively. Moreover, the FFA release rate of pH 11.0-SOB emulsions increased with increases in the heating temperatures in a period of 0 to 40 min. These results were consistent with those of Chen et al. (2022), who found that the FFA release of unheated camellia OBs was only 41.7%, while boiling treatment for 60 min increased the FFA release to 57.4%, as well as the rate of lipolysis, during the initial digestion stage [12]. Therefore, the extraction pH affected lipolysis, and the heating temperatures promoted FFA release during the initial digestion stage and had more effects on the lipolysis of pH 11.0-SOB emulsions.

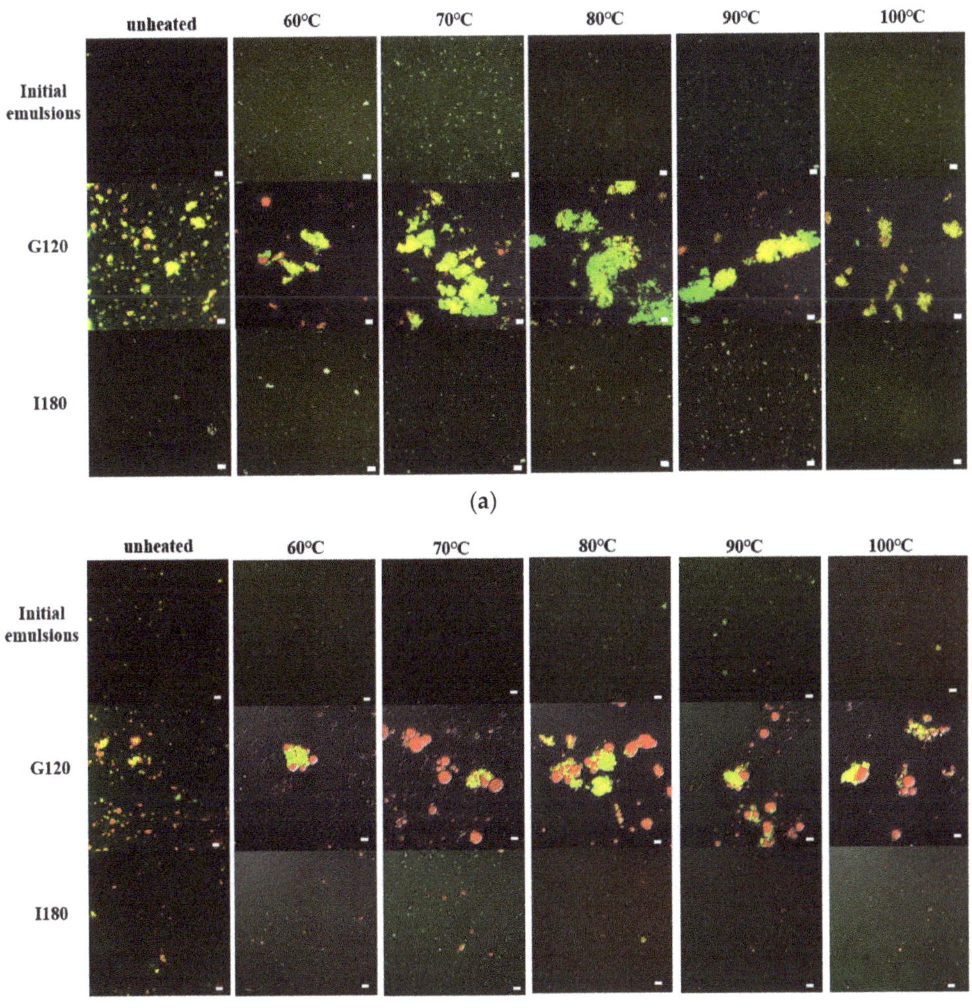


(b)

**Figure 7.** Confocal scanning microscopic images of pH 7.0-SOB (**a**) and 11.0-SOB (**b**) emulsions treated using different treatments during gastrointestinal digestion. Oil was stained using Nile Red (red), and the proteins were stained using Nile Blue (green).

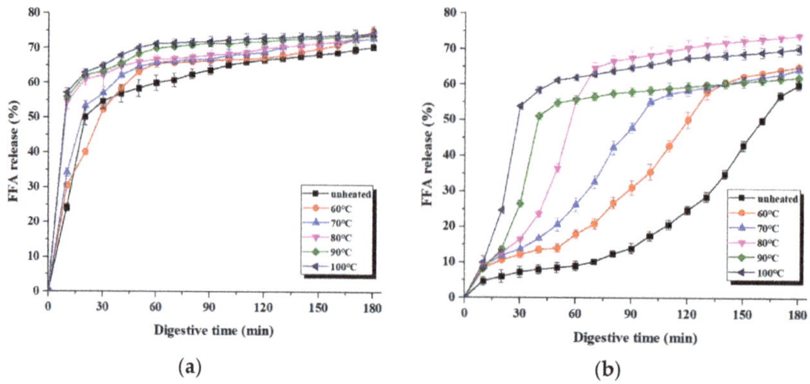


**Figure 8.** The FFA release of pH 7.0-SOB (**a**) and 11.0-SOB (**b**) emulsions treated using different treatments during intestinal digestion.

## 4. Conclusions

In this study, we prepared emulsions using SOBs extracted at pHs of 7.0 and 11.0 and discussed the effects of heat treatment (60–100 °C) on their digestive characteristics. It was found that the pH 11.0-SOB had only oleosins (16 kDa, 18 kDa, and 24 kDa), while the pH 7.0-SOB had extrinsic proteins (7S and 11S) and oleosins. Obviously, extrinsic proteins were released from pH 7.0-SOBs heated at 60–100 °C due to heat-induced denaturation. The total proportion of α-helix and β-sheet in the unheated pH 7.0-SOB was 77%, while the total proportion of α-helix and β-sheet decreased from 61.1 (60 °C) to 36.2% (100 °C). The heating temperatures had no significant effects on particle sizes and microstructures, and they had a greater effect on the zeta potentials of pH 7.0-SOBs than those of pH 11.0-SOBs before digestion. In stomach digestion, (1) the extrinsic protein hydrolysis of pH 7.0-SOB emulsions gradually accelerated from 60 to 80 °C, before slowing from 90 to 100 °C, and heat treatment inhibited oleosin hydrolysis for both the pH 7.0- and 11.0-SOB. Next, (2) the particle sizes increased for all emulsions, and heat treatment resulted in larger oil droplets in pH 11.0-SOB emulsions and larger aggregates in pH 7.0-SOB emulsions. Finally, (3) heat treatment decreased and increased the positive charges of SOB emulsions extracted at pH 7.0 and 11.0, respectively. After intestinal digestion, (1) the droplets gradually dispersed, and the particle sizes of all SOB emulsions decreased. Next, (2) heated pH 7.0–SOB emulsions had more negative charges than unheated emulsions, while the charge changes in pH 11.0-SOB emulsions showed the opposite trend. Finally, (3) for all pH 7.0-SOB emulsions, the total FFA releases were similar (about 70%) at 180 min, but different temperatures greatly affected the FFA releases of pH 11.0-SOB emulsions, and heating at 80 °C resulted in the highest FFA releases (74%). FFA release was improved in tandem with the increases in the heating temperatures during the initial digestion stage. Therefore, extraction pH and heat treatment had important effects on the digestive characteristics of SOB emulsions, which will provide guidance for selecting proper extraction pHs and heating temperatures for the application of SOB emulsions in the field of food.

**Author Contributions:** Conceptualization, D.Z. and L.Z. (Luping Zhao); Methodology, X.Y.; Software, X.Y.; Validation, L.Z. (Luyao Zhou) and Y.W.; Formal analysis, L.Z. (Luyao Zhou), Y.W. and L.Z. (Luping Zhao); Investigation, X.Y.; Resources, X.D., W.W. and D.Z.; Data curation, X.D., W.W. and L.Z. (Luping Zhao); Writing—original draft, X.Y.; Writing—review and editing, L.Z. (Luping Zhao); Visualization, L.Z. (Luyao Zhou), Y.W. and W.W.; Supervision, X.D. and D.Z.; Project administration, L.Z. (Luping Zhao); Funding acquisition, L.Z. (Luping Zhao). All authors have read and agreed to the published version of the manuscript.

**Funding:** This work was financially supported by the Natural Science Foundation of Shandong Province, PR China (No. ZR2022QC189); the Agricultural Variety Improvement Project of Shandong Province, PR China (No. 2021LZGC003); and the Young and Innovative Talents Introduction and Education Program of Universities in Shandong Province, PR China.

**Data Availability Statement:** The datasets generated for this study are available on request to the corresponding author.

**Conflicts of Interest:** There is no conflict of interest to declare.

## References

1. Zaaboul, F.; Zhao, Q.; Xu, Y.; Liu, Y. Soybean oil bodies: A review on composition, properties, food applications, and future research aspects. *Food Hydrocoll.* **2022**, *124*, 1097296. [CrossRef]
2. Ding, J.; Xu, Z.; Qi, B.; Cui, S.; Wang, T.; Jiang, L.; Zhang, Y.; Sui, X. Fabrication and characterization of soybean oil bodies encapsulated in maltodextrin and chitosan-EGCG conjugates: An in vitro digestibility study. *Food Hydrocoll.* **2019**, *94*, 519–527. [CrossRef]
3. Liu, C.; Wang, R.; He, S.; Cheng, C.; Ma, Y. The stability and gastro-intestinal digestion of curcumin emulsion stabilized with soybean oil bodies. *LWT-Food Sci. Technol.* **2020**, *131*, 109663. [CrossRef]
4. Chen, Y.; Cao, Y.; Zhao, L.; Kong, X.; Hua, Y. Macronutrients and micronutrients of soybean oil bodies extracted at different pH. *J. Food Sci.* **2014**, *79*, 1285–1291. [CrossRef] [PubMed]
5. Lin, L.; Liao, P.; Yang, H.; Tzen, J.T.C. Determination and analyses of the N-termini of oil-body proteins, steroleosin, caleosin and oleosin. *Plant Physiol. Biochem.* **2005**, *43*, 770–776. [CrossRef]
6. Cao, Y.; Zhao, L.; Ying, Y.; Kong, X.; Hua, Y.; Chen, Y. The characterization of soybean oil body integral oleosin isoforms and the effects of alkaline pH on them. *Food Chem.* **2015**, *177*, 288–294. [CrossRef]
7. Gallier, S.; Gordon, K.C.; Singh, H. Chemical and structural characterisation of almond oil bodies and bovine milk fat globules. *Food Chem.* **2012**, *132*, 1996–2006. [CrossRef]
8. Fisk, I.D.; White, D.A.; Carvalho, A.; Gray, D.A. Tocopherol—An intrinsic component of sunflower seed oil bodies. *J. Am. Oil Chem. Soc.* **2006**, *83*, 341–344. [CrossRef]
9. Groopman, E.E.; Carmody, R.N.; Wrangham, R.W. Cooking increases net energy gain from a lipid-rich food. *Am. J. Phys. Anthropol.* **2015**, *156*, 11–18. [CrossRef]
10. Yan, Z.; Zhao, L.; Kong, X.; Hua, Y.; Chen, Y. Behaviors of particle size and bound proteins of oil bodies in soymilk processing. *Food Chem.* **2016**, *194*, 881–890. [CrossRef]
11. Han, H.; Zhao, L.; Liu, X.; Guo, A.; Li, X. Effect of water bath-assisted water extraction on physical and chemical properties of soybean oil body emulsion. *Food Sci. Nutr.* **2020**, *8*, 6380–6391. [CrossRef] [PubMed]
12. Chen, C.; Pan, Y.; Niu, Y.; Peng, D.; Huang, W.; Shen, W.; Jin, W.; Huang, Q. Modulating interfacial structure and lipid digestion of natural Camellia oil body by roasting and boiling processes. *Food Chem.* **2022**, *402*, 134198. [CrossRef] [PubMed]
13. Fu, L.; He, Z.; Zeng, M.; Qin, F.; Chen, J. Effects of preheat treatments on the composition, rheological properties, and physical stability of soybean oil bodies. *J. Food Sci.* **2020**, *85*, 3150–3159. [CrossRef] [PubMed]
14. Huang, A.H.C. Oleosins and oil bodies in seeds and other organs. *Plant Physiol.* **1996**, *110*, 1055–1061. [CrossRef]
15. Capuano, E.; Pellegrini, N.; Ntone, E.; Nikiforidis, C.V. In vitro lipid digestion in raw and roasted hazelnut particles and oil bodies. *Food Funct.* **2018**, *9*, 2508–2516. [CrossRef]
16. Ding, J.; Wen, J.; Wang, J.; Tian, R.; Yu, L.; Jiang, L.; Zhang, Y.; Sui, X. The physicochemical properties and gastrointestinal fate of oleosomes from non-heated and heated soymilk. *Food Hydrocoll.* **2020**, *100*, 105418. [CrossRef]
17. Ye, A.; Wang, X.; Lin, Q.; Han, J.; Singh, H. Dynamic gastric stability and in vitro lipid digestion of whey-protein-stabilised emulsions: Effect of heat treatment. *Food Chem.* **2020**, *318*, 126463. [CrossRef]
18. Yang, X.; Wu, Y.; Liu, Y.; Ding, X.; Zhang, D.; Zhao, L. Digestive characteristics of oil body extracted from soybean aqueous extract at different pHs. *Food Res. Int.* **2022**, *161*, 111828. [CrossRef]
19. He, S.; Liu, C.; Wang, R.; Zhou, S.; Wang, Y. Comparison of two different natural oil body emulsions: In vitro gastrointestinal digestion. *J. Oleo Sci.* **2020**, *69*, 1609–1618. [CrossRef]
20. Wu, N.; Yang, X.; Teng, Z.; Yin, S.; Zhu, J.; Qi, J. Stabilization of soybean oil body emulsions using κ, ι, λ-carrageenan at different pH values. *Food Res. Int.* **2011**, *44*, 1059–1068. [CrossRef]
21. Zhao, L.; Chen, Y.; Yan, Z.; Kong, X.; Hua, Y. Physicochemical and rheological properties and oxidative stability of oil bodies recovered from soybean aqueous extract at different pHs. *Food Hydrocoll.* **2016**, *61*, 685–694. [CrossRef]
22. He, S.; Zhou, S.; Guo, W.; Wang, Y.; Liu, C.; Wang, R.; Xiao, F. Investigation of curcumin emulsion stability and gastrointestinal digestion prepared with rapeseed oil body. *J. Food Process Eng.* **2020**, *43*, 13566. [CrossRef]
23. Wu, N.; Huang, X.; Yang, X.; Guo, J.; Yin, S.; He, X.; Wang, L.; Zhu, J.; Qi, J.; Zheng, E. In vitro assessment of the bioaccessibility of fatty acids and tocopherol from soybean oil body emulsions stabilized with ι-carrageenan. *J. Agric. Food Chem.* **2012**, *60*, 1567–1575. [CrossRef]
24. Sun, Y.; Zhong, M.; Wu, L.; Wang, Q.; Li, Y.; Qi, B. Loading natural emulsions with nutraceuticals by ultrasonication: Formation and digestion properties of curcumin-loaded soybean oil bodies. *Food Hydrocoll.* **2022**, *124*, 107292. [CrossRef]

25. Gallier, S.; Tate, H.; Singh, H. In vitro gastric and intestinal digestion of a walnut oil body dispersion. *J. Agric. Food Chem.* **2013**, *61*, 410–417. [CrossRef]
26. Ono, T.; Katho, S.; Mothizuki, K. Influences of calcium and pH on protein solubility in soybean milk. *Biosci. Biotech. Biochem.* **1993**, *57*, 24–28. [CrossRef]
27. Zhao, H.; Li, W.; Qin, F.; Chen, J. Calcium sulphate-induced soya bean protein tofu-type gels: Influence of denaturation and particle size. *Int. J. Food Sci. Technol.* **2016**, *51*, 731–741. [CrossRef]
28. Zhou, X.; Sun, R.; Zhao, J.; Liu, Z.; Wang, M.; Wang, K.; Jiang, L. Enzymatic activity and stability of soybean oil body emulsions recovered under neutral and alkaline conditions: Impacts of thermal treatments. *LWT-Food Sci. Technol.* **2022**, *153*, 112545. [CrossRef]
29. McClements, D.J.; Li, Y. Review of in vitro digestion models for rapid screening of emulsion-based systems. *Food Funct.* **2010**, *1*, 32–59. [CrossRef]
30. Iwanaga, D.; Gray, D.A.; Fisk, I.D.; Decker, E.A.; Weiss, J.; McClements, D.J. Extraction and characterization of oil bodies from soybeans: A natural source of pre-emulsified soybean oil. *J. Agric. Food Chem.* **2007**, *55*, 8711–8716. [CrossRef]
31. Chen, Y.; Li, H.; Zhang, C.; Kong, X.; Hua, Y. Novel strategy for the demulsification of isolated sesame oil bodies by endogenous proteases. *J. Am. Oil Chem. Soc.* **2021**, *98*, 1057–1068. [CrossRef]

foods

MDPI

Article

# Characterization and Antibacterial Properties of Egg White Protein Films Loaded with ε-Polylysine: Evaluation of Their Degradability and Application

Xin Li [1,*], Jianhao Lv [1], Minghao Niu [1], Siqi Liu [1], Yue Wu [1], Jiahan Liu [1], Jingwen Xie [1], Chengfeng Sun [1] and Yue-Meng Wang [2,*]

1 School of Life Sciences, Yantai University, Yantai 264005, China; ljh17853509809@163.com (J.L.); nmh@ytu.edu.cn (M.N.); sageliu_xxx@163.com (S.L.); w2573630513@163.com (Y.W.); jhoney19980312@163.com (J.L.); 18763449912@163.com (J.X.); yanjunyangx@126.com (C.S.)

2 School of Food and Biological Engineering, Yantai Institute of Technology, Yantai 264003, China

* Correspondence: xinli@ytu.edu.cn (X.L.); wangyuemeng@yitsd.edu.cn (Y.-M.W.)

**Abstract:** There is an ongoing trend to design new kinds of food packaging materials with excellent properties which are environmentally friendly enough. The aim of this study was to prepare and characterize egg white protein (EWP)-based composite films with and without ε-polylysine (Lys), as well as to compare their physical–chemical properties, structural properties, degradation and antibacterial properties. The results showed that with the addition of Lys, the composite films showed a decreasing tendency of the water permeability due to the enhanced interaction between proteins and water molecules. As indicated by the structural properties, stronger cross-linking and intermolecular interactions happened with increasing concentration of Lys. In addition, the composite films presented excellent antibacterial activities against *Escherichia coli* and *Staphylococcus aureus* on chilled pork in the presence of Lys. Therefore, our prepared films might be used as a freshness-keeping material with an application in meat preservation. The biodegradation evaluation demonstrated that the composite films were environmental-friendly and have potential applications in the field of food packaging.

**Keywords:** bioactive film; antibacterial properties; ecologically friendly; application in chilled pork

## 1. Introduction

Nowadays, plastics are widely used in various fields, such as agriculture, industry, construction, packaging, cutting-edge defense industries and people's daily lives [1]. The widespread use of plastics has caused irreversible harm to the environment on which people rely for survival [2,3]. It is known that traditional plastics are mainly made of high-molecular-weight compounds such as polystyrene, polypropylene and polyvinyl chloride. The chemical structure of these polymer compounds is stable, making them difficult to degrade by microorganisms. It is reported that the degradation of traditional plastics in the natural environment takes 200 to 400 years [4,5]. Therefore, arbitrary disposal of traditional plastics is extremely prone to environmental pollution. In order to solve the problem, developing a biodegradable functional film has become a key research project in the field of food packaging [6,7].

Egg white protein (EWP) is a kind of full-price high-quality protein with a wide range of sources and has good gel properties which give EWP a unique advantage as a film-forming matrix [8]. However, the application of pure EWP films in actual packaging processes is greatly limited due to their disadvantages, such as poor water vapor permeability (WVP), easy nourishment of microorganisms, etc. [9]. Therefore, using EWP as a matrix and combining it with other polymers to improve the performance of EWP-based films has become a new research hotspot. Studies have shown that composite films prepared by mixing EWP with k-carrageenan exhibit a decrease in WVP compared to the EWP film without k-carrageenan [10]. However, the enhancement of functionality requires seeking an ideal additive for compounding. In the past few years, various bacteriostatic agents have been incorporated into biological macromolecular films to enhance their antibacterial properties as well as their practical applications [11–15]. Ma Yuan's team studied and prepared a konjac glucomannan (KGM)/chitosan (CTS) antibacterial film with oregano essential oil. The analysis of antibacterial properties showed that the addition of microcapsules improved the antibacterial performance of the KGM/CTS film and showed a better freshness during the storage of cold fresh pork when packed with the composite film [16]. However, there are currently not many reports on improving the antibacterial properties of egg white protein films. The potential application of EWP-based films in food industries still needs further research.

Polylysine (Lys) is an efficient, safe and environmentally friendly natural preservative. It is non-toxic and biodegradable and can be widely used for food preservation. Meanwhile, Lys can effectively inhibit the growth of spoilage bacteria, maintain the freshness of the product and, thus, extend the shelf life of the product. Ma, Yuan, Feng, Wang, Sun, Cao and Huang [17] reported that Lys can significantly ameliorate the gel-forming ability of meat proteins. In addition, Lys could be incorporated with carboxymethyl cellulose to increase the antibacterial activity of the composited films and could be a potential material for food packaging [18]. Therefore, both biodegradability and antibacterial properties could be guaranteed in biomacromolecule-based composite films by the addition of Lys.

The purpose of this experiment is to prepare EWP-Lys composite films and determine the physical and chemical properties, degradation ability and antibacterial effect of the films during the preservation of cold fresh meat and clarify their characteristics.

## 2. Materials and Methods

### 2.1. Chemicals

Egg white protein powder, which is greater than 95% purity, was kindly donated by Kangde Egg Industry Co., Ltd. (Nantong, China). Glycerin and ε-polylysine were purchased from Sinopharm Chemical Reagent Co., Ltd. Foodborne pathogens (*Escherichia coli* and *Staphylococcus aureus*) were kindly donated by the Bacterial Fungus Research Center in Yantai University. Pork tenderloin originated from New World Department Store (Yantai, China).

### 2.2. Film Formation Procedure

The composite films were prepared based on the previous literature [19] using the solvent casting method. Briefly, the film-forming EWP solutions (8.0% protein concentration, $w/w$) prepared with egg white powder were mixed with different concentrations of Lys (0%, 0.4%, 0.8%, 1.2%, 1.6% and 2.0%, $w/w$). The solutions were stirred for 20 min and mixed with glycerol (2.0%, $w/v$), heated in a water bath at 60 °C for 30 min and cast onto glass plates (diameter 12.5 cm). Then, the glass plates containing samples were put into the oven to dry (60 °C, 3 h). After drying, the films were peeled off and equilibrated at 25 °C ± 2 °C and 52% ± 2% relative humidity (RH) for at least 48 h before further analyses.

### 2.3. Physicochemical Characterization of the Composite Films

#### 2.3.1. Moisture content (MC), Swelling Capacity (SC), Water Solubility (WS) and Water Vapor Permeability (WVP)

Square films (2 cm × 2 cm, weight as $m_0$) were dried to a constant weight in a laboratory oven at 105 °C for 24 h and weighed as $m_1$. The SC of the films was determined by immersing them in 40 mL deionized water for 24 h at ambient temperature. The above-mentioned samples were clamped out. Subsequently, the remaining water on the surface was absorbed by filter paper. Then, the films were weighed as $m_2$. Afterward, the films were dried at 105 °C for another 24 h, and the weight was recorded as $m_3$. Therefore, the MC, SC and WS were calculated using Equations (1), (2) and (3), respectively.

$$MC(\%) = \frac{m_0 - m_1}{m_0} \times 100\% \quad (1)$$

$$SC(\%) = \frac{m_3 - m_1}{m_1} \times 100\% \quad (2)$$

$$WS(\%) = \frac{m_1 - m_2}{m_1} \times 100\% \quad (3)$$

WVP was examined using the gravimetric method, as described by Wang, Liu, Ye, Wang and Li [20], with some modifications. In the experiment of this study, a paraffin seal was employed between the composite film and the mouth of a suitable testing cup, which contained 5 g of anhydrous calcium chloride. Subsequently, the testing cups were transferred into a desiccator (75% RH), supplemented with the saturated sodium chloride solution and kept for 24 h. The WVP of the composite film was examined at 25 °C and expressed as Equation (4).

$$WVP = \frac{\Delta w \times L}{t \times A \times \Delta p} \quad (4)$$

where $t$ denotes the time (s); $A$ represents the measuring area ($m^2$); $\Delta p$ expresses the water vapor pressure difference across the two sides of the film; $\Delta w$ is the weight change of the cup (g); and $L$ expresses the thickness of the film (m).

2.3.2. Light Permeability

$L^*$, $a^*$, $b^*$ and $\Delta E$ were assessed using a colorimeter (CS-800, 400–700n, Hangzhou, China) to determine the color of the composite films. Parameters were recorded randomly for six points selected on the surfaces of each film.

Light transmittance and opacity were examined using a UV–visible spectrometer (T6, UV/Vis spectrometer, PG Instruments Ltd., Shanghai, China). Films were cut into 10 mm × 30 mm rectangles and placed on a quartz cell for determination. The respective measurement was repeated three times, and the average was taken. The opacity and transparency (T) of the composite films were calculated as follows:

$$Opcity = \frac{A_{600}}{d} \quad (5)$$

where $A_{600}$ denotes the absorbance value of the composite film at 600 nm; $d$ represents the thickness of the composite film (mm).

$$T(\%) = 0.1^A \times 100\% \quad (6)$$

where $T$ represents the transparency of the composite film; $A$ represents the absorbance of the composite film.

### 2.4. *Structural Properties of the Composite Films*

2.4.1. SEM

Cross-sections of the composite films were observed using a scanning electron microscope (SEM) (JEC 3000FC, JEOL, Tokyo, Japan). Before observation, samples were immersed in liquid nitrogen, cryo-fractured and gold-plated. Afterward, they were transferred to the SEM chamber and observed at an acceleration voltage of 20 kV.

2.4.2. FTIR

The chain interactions into films were analyzed through FTIR spectroscopy with an IS50 spectrometer (ThermoNicolet, Madison, WI, USA). The spectra were collected over a wavenumber range of 4000–400 cm$^{-1}$ and then analyzed using OMNIC Spectra software (Bruker Optics, Champs-sur-Marne, France).

### 2.5. Antibacterial Potential of the Composite Films

Antimicrobial activities of the composite films were assessed with a reported liquid culture assay. *Escherichia coli* (*E. coli*) and *Staphylococcus aureus* (*S. aureus*) were considered the main foodborne pathogens in this study. The LB liquid medium was prepared and placed in an autoclave for 20 min at 121 °C. Subsequently, the container was taken out and cooled down till the temperature was suitable for the medium to be used. The liquid incubation test was performed by cutting the films into squares (2 × 2 cm$^2$). Films were immersed into 20 mL nutrient broth supplemented with 100 μL culture of bacteria (10$^5$ CFU/mL) and then incubated at 37 °C. During the incubation, the samples (1 mL) were extracted periodically at 24 h and 48 h. Their antibacterial properties were examined by testing the optical densities using a spectrophotometer (T6, 325–1100n, Beijing, China) at 560 nm.

### 2.6. Degradation of the Composite Films

The prepared films were cut into squares (5 cm × 5 cm) and then buried in natural soil at a depth of 5.0 cm to determine their biodegradability. Subsequently, ordinary polyethylene films were also prepared to compare with the composite films as the control. The composite films and the polyethylene films were dug out and then recorded in pictures to assess their biodegradability on the 2nd day, 4th day, 6th day, 8th day, 10th day and 12th day.

### 2.7. Application of the Composite Films

Fresh pork was cut into 30 g and wrapped up in the films (with 1.2 wt% Lys as an example) and then stored in a refrigerator at 4 °C to confirm the antibacterial properties of the composite films. Meanwhile, food-grade polyethylene plastic bags served as the control group to determine the freshness of the meat. During the storage of the chilled pork, images were captured to record changes over 7 days.

### 2.8. Statistical Analysis

Data are expressed as mean value ± standard deviation of three independent experiments unless mentioned otherwise. The significance level of the differences between means was assessed at $p < 0.05$. SPSS software version 20 was used to analyze the results using a one-way analysis of variance (ANOVA) coupled with Tukey's test and reported with Origin 2019b.

## 3. Results and Discussion

### 3.1. WS, MC, SC and WVP of the Composite Film

Water solubility (WS), moisture content (MC), swelling capacity (SC) and water vapor permeability (WVP) are the key indicators of water sensitivity of protein-based films and can be used to measure their water permeability. The rapid loss of moisture in low-humidity environments can reduce the mechanical properties of the film, leading to film breakage or shrinkage. As a non-toxic and edible plasticizer, glycerol plays a role in promoting water retention and reducing water loss in the preparation of protein-based composite films. Therefore, the EWP film with and without Lys was prepared with glycerol as the plasticizer, and WS, MC, SC and WVP are shown in Table 1. As can be seen, the WS of the composite film gradually decreased with the increase in Lys addition, from 36.68% ± 2.78% to 26.45% ± 3.15%. The decrease in WS may be caused by the direct interaction between the weakly polar molecules of the EWP and Lys [21]. In addition, SC also showed a downward trend from 4.23% ± 0.36% to 2.84% ± 0.16% ($p < 0.05$). Meanwhile, molecules in the composite films will bind more tightly, which can also reduce the free volume of the composite film matrix, leading to a further decrease in SC. In addition, MC decreased from 12.11% ± 0.79% to 6.33% ± 0.05%. This was due to the increased proportion of Lys in the composite film and the enhanced interaction between proteins and water molecules, resulting in a decrease in the moisture content free in the film substrate and a decrease in MC value [22]. Thus, the addition of Lys has a significant impact on the water permeability of the composite film.

**Table 1.** Water solubility (WS), moisture content (MC), swelling capacity (SC) and water vapor permeability (WVP) of the composite films with different concentrations of Lys.

| Lys (%) | WS (%) | MC (%) | SC (%) | WVP ($gPa^{-1}s^{-1}m^{-1}\cdot 10^{-10}$) |
|---|---|---|---|---|
| 0 | 36.68 ± 2.78 [a] | 16.34 ± 2.13 [a] | 4.23 ± 0.36 [a] | 14.87 ± 1.78 [a] |
| 0.4 | 35.78 ± 1.51 [a] | 13.22 ± 1.39 [ab] | 4.15 ± 0.10 [ab] | 14.73 ± 0.40 [a] |
| 0.8 | 33.99 ± 1.19 [ab] | 14.39 ± 1.21 [ab] | 3.72 ± 0.19 [bc] | 14.37 ± 0.93 [a] |
| 1.2 | 33.03 ± 3.62 [ab] | 12.94 ± 2.29 [ab] | 3.63 ± 0.26 [c] | 13.53 ± 0.06 [a] |
| 1.6 | 29.77 ± 0.33 [bc] | 15.21 ± 1.34 [a] | 3.19 ± 0.08 [d] | 11.60 ± 0.66 [b] |
| 2.0 | 26.45 ± 3.15 [c] | 10.27 ± 1.29 [b] | 2.84 ± 0.16 [d] | 9.30 ± 0.66 [c] |

Values are presented as mean ± standard deviation. Different letters in the same column indicate significant differences ($p < 0.05$).

Water vapor permeability (WVP) is another indicator that reflects the water permeability of the composite films [23,24]. It is reported that packaging materials with low WVP can inhibit water loss and reduce nutritional loss while maintaining the complete appearance of the food itself to extend its storage time [25,26]. In general, the WVP value of packaging films is affected by the properties of composite films, film structure and environmental conditions. As can be seen in Table 1, the WVP value of the composite film decreased from the initial value of 14.87 ± 1.78 ($gPa^{-1}s^{-1}m^{-1}\cdot 10^{-10}$) to 9.30 ± 0.66 ($gPa^{-1}s^{-1}m^{-1}\cdot 10^{-10}$) with the addition of Lys, showing a significantly decreasing tendency of WVP of the composite film. This phenomenon was due to the increase in cross-linking reactions between Lys and other substances in the film, such as proteins and glycerol [27,28]. Furthermore, the transfer path of water molecules through the membrane substrate was more difficult after the addition of Lys [29].

### 3.2. Light Permeability of the Composite Film

The color of the EWP-Lys composite film is explained in Figure 1a. There were no significant changes in L* values and ΔE values with the increasing concentration of Lys. Overall, the b* value did not increase significantly, while the a* value increased from −0.75 ± 0.16 to −1.26 ± 0.04, indicating that the greenness value of the composite films increased significantly when the Lys addition reached 2%. The influence of Lys on the light performance of composite films is also manifested in its impact on optical properties. Similarly to the changes in color differences, it can be seen from Figure 1b that the concentration of Lys has no significant impact on the opacity and light transmittance of the EWP-based composite film ($p > 0.05$). However, the opacity of the composite film presented a downward trend, and the light transmittance increased when the amount of Lys added reached 1.2%. This could be attributed to the increase in protein–protein cross-linking by the participation of Lys in the film-forming reaction, which increased the molecular matrix space and then caused an increase in the visible light transmittance [30]. However, the addition of Lys had no significant impact on the optical properties of the composite films from the overall trend.

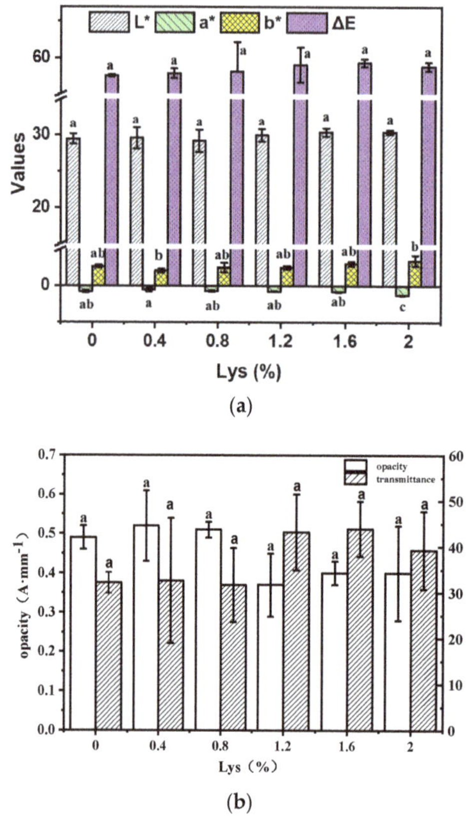


**Figure 1.** (**a**) Color parameters and (**b**) opacity and transmittance of antibacterial composite films incorporated with different concentrations of Lys (0–2.0%). Different letters indicate significant differences ($p < 0.05$).

### *3.3. Structural Properties*

#### 3.3.1. SEM Image Analysis

It has been reported that Lys, whose essence is protein, could be embedded into the macromolecular chain of EWP through protein–protein interactions, promoting the fluidity of the molecular chain and reducing the brittleness of EWP-based films [31]. Figure 2a–f represents scanning electron microscope images of EWP-Lys composite film with a concentration of Lys from 0% to 2.0% (*w*/*w*). It can be observed from Figure 2 that significant differences appeared in the cross-sectional structure and morphology of the composite films. With an increasing addition of Lys, the cross-section of the composite films gradually changed from "convolution" to "smoothness", which could be interpreted as the decrease in intermolecular force between proteins through hydrogen bonds between Lys and EWP under high-speed rotation and relatively high temperature conditions [32,33]. In addition, with the increase in the amount of Lys, the degree of cross-linking between Lys and EWP increased, and the degree of convolution on the protein-based film further decreased. At the same time, it increased the tortuosity of water molecular inlet and outlet channels, which was one of the reasons why WVP decreased with the increase in Lys. It was found by Lan, Liu, Wang, Tian, Miao, Wang and Tang [34] that the morphology of lysine/starch/PVA film was denser than that of starch/PVA film. The hydroxyl groups of starch can form hydrogen bonds with the amino groups of Lys, forming a dense structure.

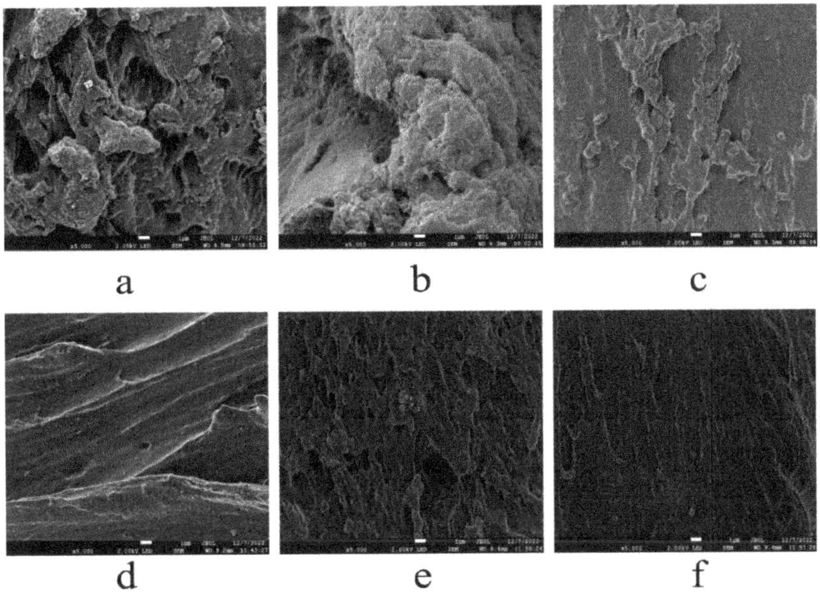

**Figure 2.** Effects of different concentrations of Lys ((**a**–**f**), 0–2%) on the cross-section structure of composite films.

3.3.2. Fourier-Transform Infrared Spectroscopy Analysis

Intermolecular interactions and changes in protein structure within the film substrate can be studied at the molecular level. As reported, molecules of Lys have hydrophilic groups (amino acid residues) that can interact with the matrix components of EWP films [35,36]. To clarify the intermolecular interactions between Lys and EWP, FTIR spectra of EWP films prepared with glycerol and different percentages of Lys are depicted in Figure 3. The characteristic peaks of EWP mainly occurred at 3404 cm$^{-1}$, 2945 cm$^{-1}$, 1639 cm$^{-1}$, 1537 cm$^{-1}$ and 1043 cm$^{-1}$, attributed to O–H and N–H stretching, C–H stretching, C=O stretching, N–H bending and C–O–C stretching vibrations, respectively [37–39]. After combining with Lys, no additional peaks of the films were observed, indicating that there was no covalent bond between EWP and Lys [40]. In other words, the interaction between the compounds is more likely due to physical reactions, and the degree of cross-linking between Lys and EWP is high enough. However, the O–H stretching vibration was depicted from 3379 cm$^{-1}$ to 3404 cm$^{-1}$ with the concentration of Lys increasing from 0% ($w/w$) to 2.0% ($w/w$), which might be attributed to the alteration of the stretching vibration in free O–H bands. In addition, C=O stretching was detected at 1662 cm$^{-1}$ and slightly shifted to 1639 cm$^{-1}$, while C–N stretching and N–H bending were observed shifting from 1539 cm$^{-1}$ to 1537 cm$^{-1}$ with the increasing addition of Lys. These phenomena were caused by C=O stretching and N–H bending in Lys in the film matrix [41]. Therefore, the spectral changes of EWP-based films should be due to alterations in the orientation and conformation of peptide chains with the incorporation of Lys.

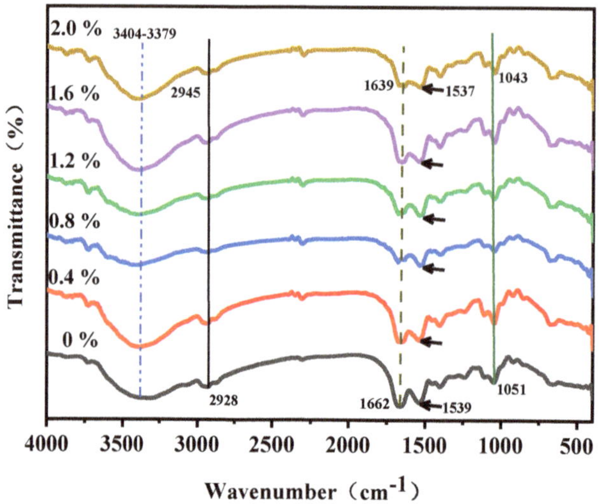


**Figure 3.** Infrared spectroscopy (FTIR) analysis of composite films with different concentrations of Lys (the black line: 0% Lys, the red line: 0.4% Lys, the blue line: 0.8% Lys, the green line: 1.2% Lys, the purple line: 1.6% Lys, the yellow line: 2.0% Lys).

### *3.4. Antibacterial Properties*

Figure 4 shows the growth of *Escherichia coli* (*E. coli*, Figure 4a) and *Staphylococcus aureus* (*Sau.*, Figure 4b) at 24 h and 48 h (represented by changes in absorbance values) after adding the antibacterial composite film with different concentrations of Lys to the liquid medium. It can be seen from Figure 4 that when the concentration of Lys reaches 2.0%, the absorbance is the lowest, and the bacteriostatic effect is the best. In addition, a stronger inhibitory effect on *E. coli* was exhibited in Figure 4a, supporting the broad-spectrum antibacterial activity of Lys [42]. The addition of Lys indeed improves the antibacterial activity of the composite film and also proves that the EWP-Lys film has a good antibacterial application prospect. The possible bacteriostatic mechanism of Lys lies in the destruction of the structure of the cell membrane of microorganisms, leading to the separation of cell walls and cytoplasm, cell collapse or cytoplasmic overflow [43,44]. Secondly, the normal physiological metabolism of bacteria could be disrupted, causing disruptions in cell material, energy and information transmission [45,46]. Thirdly, Lys can interact with internal substances in cells to destroy the cell core, resulting in the overflow of ultraviolet absorbents and ultimately leading to the death of the cell [47]. However, in-depth studies should be conducted for more information to determine the exact bacteriostatic mechanism of the composite films.

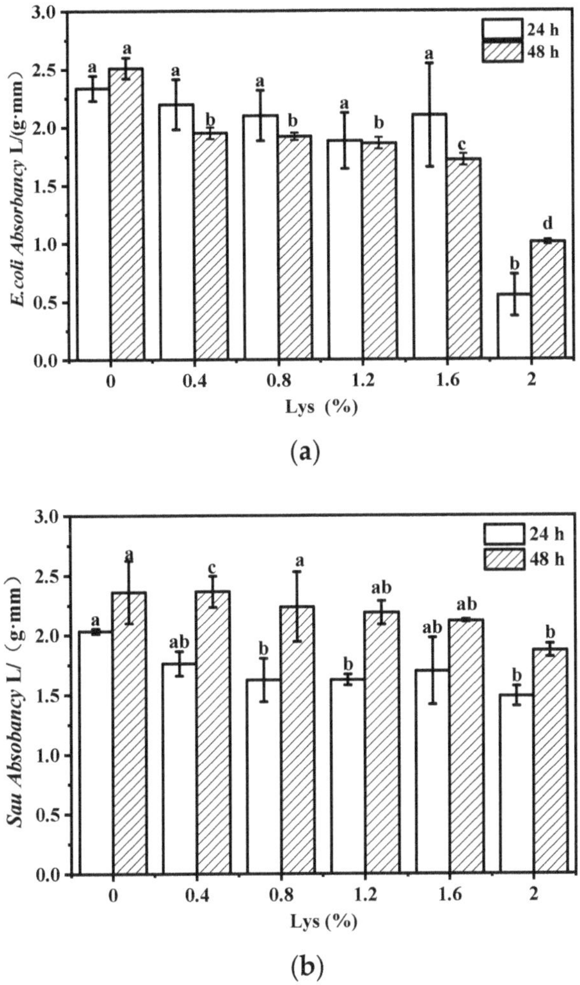


**Figure 4.** The antibacterial effects of antibacterial films on *Escherichia coli* (*E. coli*) (**a**) and *Staphylococcus aureus* (*Sau.*) (**b**) within 24 h and 48 h. Different letters indicate significant differences ($p < 0.05$).

### 3.5. Biodegradability

The biodegradability of the EWP-Lys composite film with the traditional plastic (polyethylene, PE) film as a control group is shown in Figure 5. Both groups were buried in the soil for several days, and the degradation within 12 days was observed and compared. The results showed that no signs of degradation could be seen within 12 days for the control group, while the EWP-Lys composite film exhibited degradation phenomena starting from the 4th day with wrinkles and structural defects. As time increased, the EWP-Lys group gradually fractured and lost and degraded into fragments of the size of soybean seeds by the 12th day. However, the size and morphology of the control group did not change at all. Therefore, the biodegradability of EWP-Lys films is high enough, and it is estimated that it will take about two weeks for completely degradation. The excellent degradation of EWP-Lys composite film provides great potential for its subsequent practical application, without any concern about environmental damage [48].

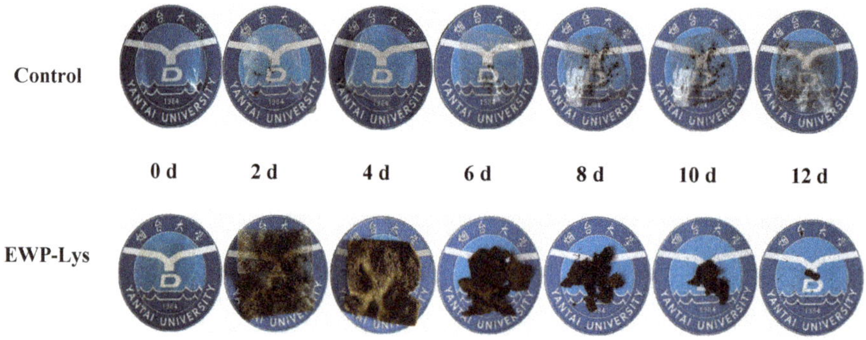


**Figure 5.** Comparison of degradation effects of polyethylene plastic film (control) and the composite film (EWP-Lys) with 1.2% Lys.

### 3.6. Application

#### 3.6.1. pH Values of the Chilled Pork during Storage

Meat products are prone to microbial contamination during storage due to their rich protein content. Bacterial microorganisms are extremely prone to decomposing proteins into ammonia substances, resulting in an increase in pH [49]. Figure 6 shows the change in pH of the chilled pork during its storage period when wrapped in the composite film (with 1.2 wt% Lys), while the chilled pork wrapped in PE film was used as the control group. As shown, the pH of the Lys group on the first and seventh days of storage is slightly higher than that of the control group. This is because a small portion of the protein present in the composite film might adhere to the chilled pork, and the protein is oxidized and decomposed to release alkaline groups at the beginning of storage, resulting in a higher pH than that of the control group. On the seventh day, the water molecular channel of the composite film is opened due to protein decomposition, resulting in an increase in the pH in the environment. Considering the actual storage situation, when this composite film is applied to the preservation of cold fresh meat, the recommended storage time is 3–5 days, and the preservation effect is slightly higher than that of ordinary polyethylene film.

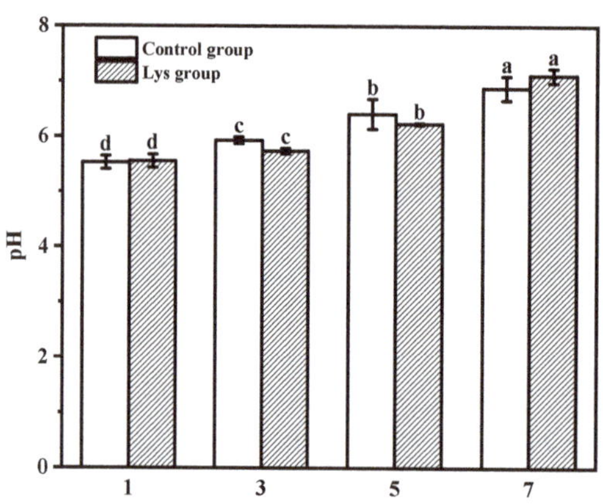


**Figure 6.** Changes in the pH value of the chilled meat wrapped in polyethylene plastic film (control group) and the composite film (Lys group, with 1.2 wt% Lys) during storage (mean ± standard error). Different letters indicate significant differences ($p < 0.05$).

### 3.6.2. Color of the Chilled Pork during Storage

The color of chilled meat is the most intuitive method used by consumers to evaluate the freshness of meat products [50]. The higher the redness value (a*) of displayed meat, the more popular it is with consumers. Therefore, the a* value is a very important indicator in the pork color evaluation system. Table 2 shows the impact of two types of packaging (i.e., PE film and EWP-Lys film) on the color value of chilled pork. As can be seen, the brightness value (L*) of chilled pork shows an upward trend with the extension of storage time, while the yellowness value (b*) is not significantly affected during the whole storage period. The redness value (a*) shows a downward trend with the prolongation of storage time, mainly due to the gradual increase in the degree of lipid oxidation of the chilled pork and the color change caused by water loss over time. In the early stage of storage, the a* value of chilled pork in different packages changed slowly. After storing for 3 days, the a* value of the composite film increased and was significantly lower than that of the control group (PE film) ($p < 0.05$), indicating that the storage water loss rate of the chilled pork wrapped in PE film was greater than that of the experimental group. Thus, the composite film has a good preservation effect in practical applications.

**Table 2.** Changes in the Chrominance value of the chilled meat during storage (mean ± standard error).

| Chromatic Aberration | Storage Time (d) | Control Group | Lys Group |
|---|---|---|---|
| L* | 1 | $55.69 \pm 1.11^{c}$ | $42.16 \pm 0.79^{b}$ |
| | 3 | $59.35 \pm 0.54^{b}$ | $41.03 \pm 1.89^{b}$ |
| | 5 | $61.71 \pm 1.06^{a}$ | $46.37 \pm 0.50^{a}$ |
| | 7 | $61.12 \pm 0.06^{a}$ | $46.54 \pm 2.92^{a}$ |
| a* | 1 | $3.07 \pm 0.40^{a}$ | $2.07 \pm 0.34^{b}$ |
| | 3 | $2.05 \pm 0.30^{b}$ | $3.47 \pm 0.41^{a}$ |
| | 5 | $0.89 \pm 0.08^{c}$ | $1.87 \pm 0.03^{b}$ |
| | 7 | $0.48 \pm 0.04^{d}$ | $1.46 \pm 0.06^{c}$ |
| b* | 1 | $12.82 \pm 0.45^{a}$ | $10.57 \pm 0.85^{a}$ |
| | 3 | $8.06 \pm 0.82^{c}$ | $9.86 \pm 1.32^{a}$ |
| | 5 | $11.58 \pm 0.61^{b}$ | $8.39 \pm 0.09^{b}$ |
| | 7 | $10.90 \pm 0.04^{b}$ | $10.33 \pm 0.44^{a}$ |
| Δ E | 1 | $34.68 \pm 1.03^{a}$ | $47.38 \pm 0.65^{a}$ |
| | 3 | $30.49 \pm 0.54^{b}$ | $48.95 \pm 1.80^{a}$ |
| | 5 | $29.46 \pm 0.86^{b}$ | $43.79 \pm 0.17^{b}$ |
| | 7 | $27.48 \pm 1.89^{c}$ | $41.04 \pm 3.95^{b}$ |

Values are presented as mean ± standard deviation. Different letters in the same column indicate significant differences ($p < 0.05$).

### 3.6.3. Total Bacterial Counts of the Chilled Pork during Storage

Cold fresh meat is susceptible to microbial contamination during storage [51]. Figure 7 shows the growth of microorganisms in chilled pork wrapped in PE film and EWP-Lys composite film during the storage period. It can be clearly seen from the figure that there is a slight growth of microorganisms in the Lys group at the initial stage of storage, which is consistent with the measurement results of indicators such as the pH of chilled pork. With the increase in storage time, the microbial growth of the Lys group was significantly lower than that of the control group (PE film) after 3 days (Table 3). Lys in the composite films played an effective role in inhibiting microbial growth, which means that the bacteriostatic effect of Lys was initially effective. After the storage time was extended to 7 days, the microbial growth rate was still lower than that of the control group. This result can more intuitively demonstrate that the composite film can play a key role in the storage process of cold fresh meat.

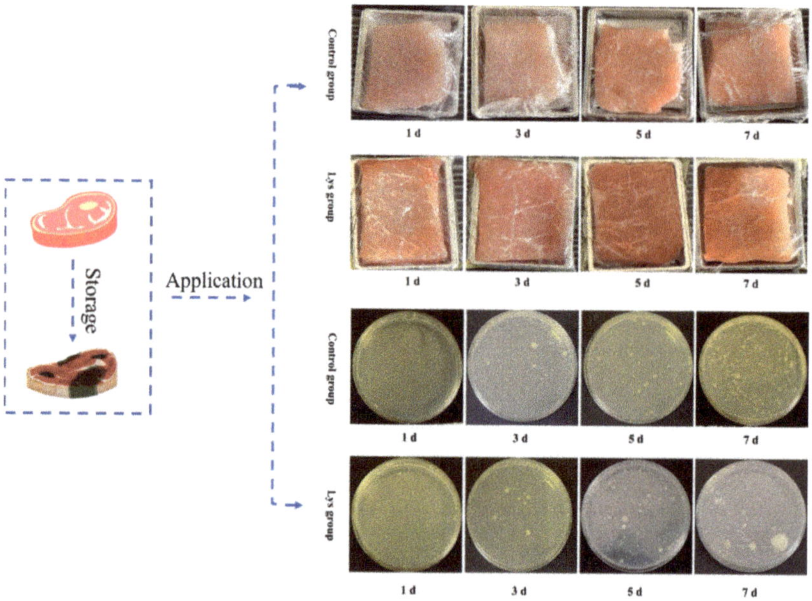


**Figure 7.** The application of the EWP-Lys films (with 1.2 wt% Lys as an example) in the preservation of chilled pork, including the apparent images and the colony changes (polyethylene film was used as a control group).

**Table 3.** The number of colonies in the chilled meat during storage (mean ± standard error).

| Group | 1 d (CFU) | 3 d (CFU) | 5 d (CFU) | 7 d (CFU) |
|---|---|---|---|---|
| Control | $(1.00 \pm 0.28) \times 10^{-4}$ | $(8.15 \pm 1.63) \times 10^{-4}$ | $(1.82 \pm 0.37) \times 10^{-5}$ | $(3.85 \pm 0.27) \times 10^{-5}$ |
| Lys | $(2.85 \pm 0.35) \times 10^{-4}$ | $(4.90 \pm 0.56) \times 10^{-4}$ | $(9.85 \pm 0.92) \times 10^{-4}$ | $(1.57 \pm 0.30) \times 10^{-5}$ |

Values are presented as mean ± standard deviation.

## 4. Conclusions

Lys is used as a bacteriostatic agent with different concentrations in EWP-based composite films to improve their physico-chemical and antibacterial properties, aiming to develop alternative and environmentally friendly composite films applicable to the food-packaging field in chilled meat. Measurements of biodegradation seemed to support the idea that the composite films were much easier to degrade than the polyethylene films, which demonstrated excellent biodegradability. Their structural properties showed that a stronger cross-linking occurred with the increasing addition of Lys. Furthermore, antibacterial effects on microorganisms have been proved and demonstrated in the application of chilled pork. Thus, the current study demonstrates that EWP-Lys composite films could be considered as an ideal option for environment-friendly food packaging, especially for chilled meat, due to the improved physico-chemical properties, antibacterial properties and degradable properties.

**Author Contributions:** X.L.: conceptualization, methodology, data curation, software, writing—review and editing, supervision and funding acquisition. J.L. (Jianhao Lv): data curation, methodology, writing and editing. M.N.: investigation, conceptualization, methodology and data curation. S.L.: investigation, conceptualization, methodology and data curation. Y.W.: investigation, conceptualization, methodology and data curation. J.L. (Jiahan Liu): investigation, conceptualization, methodology and data curation. J.X.: data curation and validation. C.S.: visualization and formal analysis. Y.-M.W.: supervision, methodology and writing—review and editing. All authors have read and agreed to the published version of the manuscript.

**Funding:** This study was carried out with support of the Youth Fund of Shandong Natural Science Foundation (Project No. ZR2021QC065) and the Research Startup Fund from Yantai University (SM20B61), as well as Yantai University Graduate Research (Technology) Innovation Fund (Project No. GGIFYTU2356). The authors would like to thank all of the individuals who volunteered for this study.

**Data Availability Statement:** The datasets generated for this study are available on request to the corresponding author.

**Conflicts of Interest:** The authors declare no conflict of interest.

## References

1. Hussain, I.; Ganiyu, S.A.; Alasiri, H.; Alhooshani, K. A state-of-the-art review on waste plastics-derived aviation fuel: Unveiling the heterogeneous catalytic systems and techno-economy feasibility of catalytic pyrolysis. *Energy Convers. Manag.* **2022**, *274*, 116433. [CrossRef]
2. Schwarz, A.; Lensen, S.; Langeveld, E.; Parker, L.; Urbanus, J. Plastics in the global environment assessed through material flow analysis, degradation and environmental transportation. *Sci. Total Environ.* **2023**, *875*, 162644. [CrossRef] [PubMed]
3. Nguyen, L.H.; Nguyen, B.-S.; Le, D.-T.; Alomar, T.S.; AlMasoud, N.; Ghotekar, S.; Oza, R.; Raizada, P.; Singh, P.; Nguyen, V.-H. A concept for the biotechnological minimizing of emerging plastics, micro- and nano-plastics pollutants from the environment: A review. *Environ. Res.* **2023**, *216*, 114342. [CrossRef]
4. Zunita, M.; Winoto, H.P.; Fauzan, M.F.K.; Haikal, R. Recent advances in plastics waste degradation using ionic liquid-based process. *Polym. Degrad. Stab.* **2023**, *211*, 110320. [CrossRef]
5. Nabi, I.; Bacha, A.-U.; Ahmad, F.; Zhang, L. Application of titanium dioxide for the photocatalytic degradation of macro- and micro-plastics: A review. *J. Environ. Chem. Eng.* **2021**, *9*, 105964. [CrossRef]
6. Kontominas, M.G. Use of Alginates as Food Packaging Materials. *Foods* **2020**, *9*, 1440. [CrossRef]
7. Tan, C.; Pang, D.; Wu, R.; Zou, F.; Zhang, B.; Shang, N.; Li, P. Development of a Multifunctional Edible Coating and Its Preservation Effect on Sturgeon (*Acipenser baeri*♀× *Acipenser schrenckii*♂) Fillets during Refrigerated Storage at 4 °C. *Foods* **2022**, *11*, 3380. [CrossRef]
8. Peng, N.; Gu, L.; Li, J.; Chang, C.; Li, X.; Su, Y.; Yang, Y. Films Based on Egg White Protein and Succinylated Casein Cross-Linked with Transglutaminase. *Food Bioprocess Technol.* **2017**, *10*, 1422–1430. [CrossRef]
9. Razi, S.M.; Fahim, H.; Amirabadi, S.; Rashidinejad, A. An overview of the functional properties of egg white proteins and their application in the food industry. *Food Hydrocoll.* **2023**, *135*, 108183. [CrossRef]
10. Huang, X.; Luo, X.; Liu, L.; Dong, K.; Yang, R.; Lin, C.; Song, H.; Li, S.; Huang, Q. Formation mechanism of egg white protein/κ-Carrageenan composite film and its application to oil packaging. *Food Hydrocoll.* **2020**, *105*, 105780. [CrossRef]
11. Xiao, Y.; Liu, Y.; Kang, S.; Wang, K.; Xu, H. Development and evaluation of soy protein isolate-based antibacterial nanocomposite films containing cellulose nanocrystals and zinc oxide nanoparticles. *Food Hydrocoll.* **2020**, *106*, 105898. [CrossRef]
12. Hu, J.; Li, D.; Huai, Q.; Geng, M.; Sun, Z.; Wang, M.; Wang, S.; Li, Y.; Zheng, H. Development and evaluation of soybean protein isolate–based antibacterial nanocomposite films containing nano-$TiO_2$. *Ind. Crops Prod.* **2023**, *197*, 116620. [CrossRef]
13. Chollakup, R.; Pongburoos, S.; Boonsong, W.; Khanoonkon, N.; Kongsin, K.; Sothornvit, R.; Sukyai, P.; Sukatta, U.; Harnkarnsujarit, N. Antioxidant and antibacterial activities of cassava starch and whey protein blend films containing rambutan peel extract and cinnamon oil for active packaging. *LWT* **2020**, *130*, 109573. [CrossRef]
14. Zou, Z.; Ismail, B.B.; Zhang, X.; Yang, Z.; Liu, D.; Guo, M. Improving barrier and antibacterial properties of chitosan composite films by incorporating lignin nanoparticles and acylated soy protein isolate nanogel. *Food Hydrocoll.* **2023**, *134*, 108091. [CrossRef]
15. Jeon, Y.J.; Lee, H.; Min, S.C. Effects of in-package atmospheric dielectric barrier discharge cold plasma treatment on the antimicrobial efficacy of whey protein isolate-based edible films that incorporate malic acid against Salmonella in chicken breast processed meat. *Innov. Food Sci. Emerg. Technol.* **2023**, *85*, 103339. [CrossRef]
16. Ma, Y.; Ye, K.; Liu, P.; Yuan, A.; Chen, S.; He, Y. Effect of a Konjac glucomannan/chitosan antibacterial composite membrane microencapsulated with oregano essential oil on the quality of chilled pork. *Appl. Food Res.* **2023**, *3*, 100249. [CrossRef]
17. Ma, W.; Yuan, F.; Feng, L.; Wang, J.; Sun, Y.; Cao, Y.; Huang, J. ε-Polylysine-mediated enhancement of the structural stability and gelling properties of myofibrillar protein under oxidative stress. *Int. J. Biol. Macromol.* **2022**, *220*, 1114–1123. [CrossRef]
18. He, B.; Wang, Y.; Jiang, Z.; Liu, S.; Zhu, J. Physical properties and antibacterial activity of the composited films based on carboxymethyl cellulose and gelatin functionalized with ε-polylysine. *Int. J. Biol. Macromol.* **2021**, *191*, 1126–1136. [CrossRef]
19. Li, X.; Liu, A.; Ye, R.; Wang, Y.; Wang, W. Fabrication of gelatin–laponite composite films: Effect of the concentration of laponite on physical properties and the freshness of meat during storage. *Food Hydrocoll.* **2015**, *44*, 390–398. [CrossRef]
20. Wang, Y.; Liu, A.; Ye, R.; Wang, W.; Li, X. Transglutaminase-induced crosslinking of gelatin–calcium carbonate composite films. *Food Chem.* **2015**, *166*, 414–422. [CrossRef]
21. Gan, L.; Jiang, G.; Yang, Y.; Zheng, B.; Zhang, S.; Li, X.; Tian, Y.; Peng, B. Development and characterization of levan/pullulan/chitosan edible films enriched with ε-polylysine for active food packaging. *Food Chem.* **2022**, *388*, 132989. [CrossRef] [PubMed]

22. Xiao, Y.; Xu, H.; Zhou, Q.; Li, W.; Gao, J.; Liao, X.; Yu, Z.; Zheng, M.; Zhou, Y.; Sui, X.; et al. Influence mechanism of wheat bran cellulose and cellulose nanocrystals on the storage stability of soy protein isolate films: Conformation modification and molecular interaction perspective. *Food Hydrocoll.* **2023**, *139*, 108475. [CrossRef]
23. Mousavi, Z.; Naseri, M.; Babaei, S.; Hosseini, S.; Shekarforoush, S. The effect of cross-linker type on structural, antimicrobial and controlled release properties of fish gelatin-chitosan composite films incorporated with epsilon-poly-l-lysine. *Int. J. Biol. Macromol.* **2021**, *183*, 1743–1752. [CrossRef] [PubMed]
24. Sáez-Orviz, S.; Rendueles, M.; Díaz, M. Impact of adding prebiotics and probiotics on the characteristics of edible films and coatings—A review. *Food Res. Int.* **2023**, *164*, 112381. [CrossRef]
25. He, Y.; Ye, H.-C.; You, T.-T.; Xu, F. Sustainable and multifunctional cellulose-lignin films with excellent antibacterial and UV-shielding for active food packaging. *Food Hydrocoll.* **2023**, *137*, 108355. [CrossRef]
26. Kumar, A.; Hasan, M.; Mangaraj, S.; Pravitha, M.; Verma, D.K.; Srivastav, P.P. Trends in Edible Packaging Films and Its Prospective Future in Food: A Review. *Appl. Food Res.* **2022**, *2*, 100118.
27. Zhang, W.; Zeng, Q.-M.; Tang, R.-C. Gallic acid functionalized polylysine for endowing cotton fiber with antibacterial, antioxidant, and drug delivery properties. *Int. J. Biol. Macromol.* **2022**, *216*, 65–74. [CrossRef]
28. Cai, L.; Cao, A.; Bai, F.; Li, J. Effect of ε-polylysine in combination with alginate coating treatment on physicochemical and microbial characteristics of Japanese sea bass (*Lateolabrax japonicas*) during refrigerated storage. *LWT Food Sci. Technol.* **2015**, *62*, 1053–1059. [CrossRef]
29. Chen, H.; Wu, C.; Feng, X.; He, M.; Zhu, X.; Li, Y.; Teng, F. Effects of two fatty acids on soy protein isolate/sodium alginate edible films: Structures and properties. *LWT* **2022**, *159*, 113221. [CrossRef]
30. Azeredo, H.M.; Waldron, K.W. Crosslinking in polysaccharide and protein films and coatings for food contact—A review. *Trends Food Sci. Technol.* **2016**, *52*, 109–122. [CrossRef]
31. Abeykoon, A.H.; Noinaj, N.; Choi, B.-E.; Wise, L.; He, Y.; Chao, C.-C.; Wang, G.; Gucek, M.; Ching, W.-M.; Chock, P.B.; et al. Structural Insights into Substrate Recognition and Catalysis in Outer Membrane Protein B (OmpB) by Protein-lysine Methyltransferases from Rickettsia. *J. Biol. Chem.* **2016**, *291*, 19962–19974. [CrossRef]
32. Lvov, Y.; Ariga, K.; Ichinose, I.; Kunitake, T. Assembly of Multicomponent Protein Films by Means of Electrostatic Layer-by-Layer Adsorption. *J. Am. Chem. Soc.* **1995**, *117*, 6117–6123. [CrossRef]
33. Cheng, J.; Li, Z.; Wang, J.; Zhu, Z.; Yi, J.; Chen, B.; Cui, L. Structural characteristics of pea protein isolate (PPI) modified by high-pressure homogenization and its relation to the packaging properties of PPI edible film. *Food Chem.* **2022**, *388*, 132974. [CrossRef] [PubMed]
34. Lan, X.; Liu, Y.; Wang, Y.; Tian, F.; Miao, X.; Wang, H.; Tang, Y. Coaxial electrospun PVA/PCL nanofibers with dual release of tea polyphenols and epsilon-poly (L-lysine) as antioxidant and antibacterial wound dressing materials. *Int. J. Pharm.* **2021**, *601*, 120525. [CrossRef]
35. Meng, Y.; Xue, Q.; Chen, J.; Li, Y.; Shao, Z. Structure, stability, rheology, and texture properties of epsilon-polylysine-whey protein complexes. *J. Dairy Sci.* **2022**, *105*, 3746–3757. [CrossRef] [PubMed]
36. Salma-Ancane, K.; Sceglovs, A.; Tracuma, E.; Wychowaniec, J.; Aunina, K.; Ramata-Stunda, A.; Nikolajeva, V.; Loca, D. Effect of crosslinking strategy on the biological, antibacterial and physicochemical performance of hyaluronic acid and ε-polylysine based hydrogels. *Int. J. Biol. Macromol.* **2022**, *208*, 995–1008. [CrossRef] [PubMed]
37. Jiang, J.; Watowita, P.; Chen, R.; Shi, Y.; Geng, J.-T.; Takahashi, K.; Li, L.; Osako, K. Multilayer gelatin/myofibrillar films containing clove essential oil: Properties, protein-phenolic interactions, and migration of active compounds. *Food Packag. Shelf Life* **2022**, *32*, 100842. [CrossRef]
38. Celebioglu, A.; Uyar, T. Electrohydrodynamic encapsulation of eugenol-cyclodextrin complexes in pullulan nanofibers. *Food Hydrocoll.* **2020**, *111*, 106264. [CrossRef]
39. Zheng, T.; Tang, P.; Yang, C.; Ran, R.; Li, G. Development of active packaging films based on collagen/gallic acid-grafted chitosan incorporating with ε-polylysine for pork preservation. *Food Hydrocoll.* **2023**, *140*, 108590. [CrossRef]
40. Sai, F.; Zhang, H.; Qu, J.; Wang, J.; Zhu, X.; Ye, P.; Zhang, Z. Thermal-driven self-healing and green recyclable waterborne polyurethane films based on double reversible covalent bonds. *Prog. Org. Coat.* **2023**, *178*, 107460. [CrossRef]
41. Zhang, L.; Wang, Z.; Jiao, Y.; Wang, Z.; Tang, X.; Du, Z.; Zhang, Z.; Lu, S.; Qiao, C.; Cui, J. Biodegradable packaging films with ε-polylysine/ZIF-L composites. *LWT* **2022**, *166*, 113776. [CrossRef]
42. He, S.; Liu, A.; Zhang, J.; Liu, J.; Shao, W. Preparation of ε-polylysine and hyaluronic acid self-assembled microspheres loaded bacterial cellulose aerogels with excellent antibacterial activity. *Colloids Surf. A Physicochem. Eng. Asp.* **2022**, *654*, 130114. [CrossRef]
43. Zhang, L.; Zhang, M.; Mujumdar, A.S.; Yu, D.; Wang, H. Potential nano bacteriostatic agents to be used in meat-based foods processing and storage: A critical review. *Trends Food Sci. Technol.* **2023**, *131*, 77–90. [CrossRef]
44. Li, Q.; Zhou, W.; Zhang, J.; Zhu, J.; Sun, T.; Li, J.; Cheng, L. Synergistic effects of ε-polylysine hydrochloride and gallic acid on *Shewanella putrefaciens* and quality of refrigerated sea bass fillets. *Food Control* **2022**, *139*, 109070. [CrossRef]
45. Wang, X.; Xie, Y.; Li, Z.; Chen, Q.; Sun, J.; Han, X.; Liu, R.; Hu, Z.; Liang, C.; Zhang, C.; et al. Honokiol inhibits growth and improves control efficiency against *Phytophthora nicotianae* through disrupting redox homeostasis and energy metabolism. *Ind. Crop. Prod.* **2022**, *178*, 114656. [CrossRef]

46. Bhattacharjee, R.; Negi, A.; Bhattacharya, B.; Dey, T.; Mitra, P.; Preetam, S.; Kumar, L.; Kar, S.; Das, S.S.; Iqbal, D.; et al. Nanotheranostics to target antibiotic-resistant bacteria: Strategies and applications. *Open Nano* **2023**, *11*, 100138. [CrossRef]
47. Wu, Y.; He, Q.; Che, X.; Liu, F.; Lu, J.; Kong, X. Effect of number of lysine motifs on the bactericidal and hemolytic activity of short cationic antimicrobial peptides. *Biochem. Biophys. Res. Commun.* **2023**, *648*, 66–71. [CrossRef]
48. Nabi, L.; Nourani, M. Biodegradable form stable phase change material for cold storage packaging of meat. *Meat Sci.* **2023**, *201*, 109188. [CrossRef]
49. Tan, C.; Hu, J.; Gao, B.; Zhang, B.; Li, P.; Shang, N. Effects of the interaction between *Aeromonas sobria* and *Macrococcus caseolyticus* on protein degradation of refrigerated sturgeon fillets: Novel perspective on fish spoilage. *LWT* **2023**, *183*, 114908. [CrossRef]
50. Yun, X.; Chen, W.; Zhang, J.; Dong, T. Colorimetric porous microspheres of natural sodium alginate for chilled pork visual monitoring. *Int. J. Biol. Macromol.* **2023**, *230*, 123198. [CrossRef]
51. Lian, F.; Jøstensen, O.; Siikavuopio, S.I.; Lorentzen, G. Live holding of red king crab (*Paralithodes camtschaticus*) and snow crab (*Chionoecetes opilio*)—Effect on microbial growth in processed leg meat during refrigerated storage. *Food Microbiol.* **2021**, *104*, 103973. [CrossRef] [PubMed]

foods

MDPI

Article

# Influence of *Cmr1* in the Regulation of Antioxidant Function Melanin Biosynthesis in *Aureobasidium pullulans*

Wan Wang [1,†], Kai Zhang [1,†], Congyu Lin [2], Shanshan Zhao [3], Jiaqi Guan [1], Wei Zhou [1], Xin Ru [1], Hua Cong [1] and Qian Yang [1,4,*]

1 School of Life Science and Technology, Harbin Institute of Technology, Harbin 150006, China; 17862702273@163.com (W.W.); zhangkaiown@163.com (K.Z.); g13610013008@163.com (J.G.); zw3268133440@163.com (W.Z.); 18745180137@163.com (X.R.); conghua@hit.edu.cn (H.C.)

2 National Engineering Research Center of Cereal Fermentation and Food Biomanufacturing, School of Food Science and Technology, Jiangnan University, Wuxi 214122, China; lincongyu@jiangnan.edu.cn

3 Ocean College, Zhejiang University, Zhoushan 316000, China; zhaoshanshan5612@163.com

4 State Key Laboratory of Urban Water Resources and Environment, Harbin Institute of Technology, Harbin 150090, China

* Correspondence: yangq@hit.edu.cn

† These authors contributed equally to this work.

**Abstract:** We have successfully identified the transcription factor *Cmr1* from the fungus *Aureobasidium pullulans* Hit-lcy3T, which regulates melanin biosynthesis genes. Bioinformatics analysis revealed that the Cmr1 gene encodes a protein of 945 amino acids, containing two $Cys_2His_2$ zinc finger domains and a $Zn(II)_2Cys_6$ binuclear cluster domain located at the N-terminus of *Cmr1*. To investigate the function of the *Cmr1* gene, we performed gene knockout and overexpression experiments. Our results showed that *Cmr1* is a key regulator of melanin synthesis in Hit-lcy3$^T$, and its absence caused developmental defects. Conversely, overexpression of *Cmr1* significantly increased the number of chlamydospores in Hit-lcy3$^T$ and improved melanin production. RT-qPCR analysis further revealed that overexpression of Cmr1 enhanced the expression of several genes involved in melanin biosynthesis, including *Cmr1*, *PKS*, *SCD1*, and *THR1*. Melanin extracted from the Hit-lcy3$^T$ was characterized using UV and IR spectroscopy. Furthermore, we assessed the antioxidant properties of Hit-lcy3$^T$ melanin and found that it possesses strong scavenging activity against DPPH·, ABTS·, and OH·, but weaker activity against $O_2^{-}$·. These findings suggest that Hit-lcy3$^T$ melanin holds promise for future development as a functional food additive.

**Keywords:** *Aureobasidium pullulans*; *Cmr1*; melanin; antioxidant activity

## 1. Introduction

Melanin is a complex and diverse large molecule named by the Swiss scientist Jöns Berzelius. It is formed by the polymerization of polyhydroxyindole or polyhydroxyphenolic substances and is widely found in animals, plants, and microorganisms [1]. Currently, melanin can be classified into five types based on the precursor substances of its biosynthetic pathway: L-DOPA (L-3,4-dihydroxyphenylalanine) melanin, 1,8-dihydroxynaphthalene (1,8-DHN) melanin, pyomelanin, pheomelanin, and Glutaminyl-4-hydroxybenzene (GHB) melanin. However, most fungi synthesize 1,8-DHN melanin and L-DOPA melanin via the DHN and L-DOPA pathways, respectively. They form complex polyphenol-based polymers through enzyme-catalyzed and chemical reactions between 1,8-DHN and L-Dopaquinone. Melanin's amorphous quality renders it insoluble in water, most organic solvents, and acidic solutions, but soluble in alkaline solutions. It is tightly bound to macromolecules such as proteins, carbohydrates, and tannins, making it difficult to accurately analyze its precise structure and chemical composition. Melanin provides protection against UV, chemical, and thermal stresses. It also acts as a potent cation chelator and antioxidant, binds and neutralizes antimicrobial peptides and antibiotics drugs, inhibits enzyme activation, and confers dehydration tolerance [2,3].

Daily foods rich in melanin include coffee, cocoa, dark beer, soy sauce, baked foods, mushrooms, and others [4]. These melanins can provide food with antioxidant capacity, inhibit oxidative stress in cells, suppress or kill pathogenic microorganisms, and induce the synthesis of chemopreventive enzymes [5]. In recent years, melanin's unique biological characteristics have led to its widespread use in a variety of fields, including food, medicine, cosmetics, and material synthesis [6,7]. Lee et al. found that melanin can effectively reduce the weight of high-fat diet mice, improve lipid status, and increase antioxidant enzyme activity to lower blood lipid levels, and water-soluble melanin complexes have been found to improve insulin sensitivity in obese mice fed a high-fat diet and effectively lower blood sugar [8]. Xu et al. verified that melanin could increase mouse inflammatory factors, alleviate inflammatory stress, and protect the liver [9]. In addition, research has shown that 20% of the total antioxidant capacity intake in the Spanish population is provided by melanin in food [10]. All of these indicate the potential application of melanin as a functional food additive [11]. The characteristics of melanin make it highly valuable, but currently, the process of extracting melanin from animals and plants is complex, with low yield and high cost, making it impossible to obtain melanin in large quantities.

*Aureobasidium pullulans* (*A. pullulans*), commonly known as black yeast, is a fascinating fungus with a complex lifecycle that exhibits various morphological forms under different environmental conditions or growth stages, including budding spores, mycelia, chlamydospores, and small oval-shaped yeast-like cells [12,13]. One of the most notable characteristics of this species is the production of melanin. Researchers have explored different approaches to optimize the production of melanin by *A. pullulans*. For example, Mujdeci used carrot peel extract as a substrate to ferment *A. pullulans* NBRC 100716 and achieved a 100% increase in the total concentration of melanin in the extract by optimizing the fermentation parameters [14]. Mujdeci found that *A. pullulans* NBRC 100716 can also synthesize natural melanin by fermenting food residues, and characterized it after extraction, with the highest yield reaching 300.0 mM/L [15]. Similarly, Gamal et al. achieved a significant increase in melanin production (up to 550.0 mM/L) by optimizing the fermentation parameters using food waste as a substrate [16]. However, most studies on melanin production by *A. pullulans* have focused on the selection of fermentation substrates and strain screening, and further research is needed to better understand the synthesis and regulation of melanin by this species.

*Cmr1* is a crucial transcriptional activator that regulates the melanin synthesis pathway in fungi. It was first discovered by Gento in 2000 and is typically composed of two $Cys_2His_2$ zinc fingers and one $Zn(II)_2Cys_6$ binuclear cluster domain [17]. Since then, homologs of *Cmr1* have been identified in many fungi, including *Helminthosporium maydis*, *Bipolaris oryzae*, *Alternaria brassicicola*, *Alternaria alternate*, *Grifola frondosa*, *Botrytis cinerea*, and *Verticillium dahliae*. These homologs play regulatory roles in the DHN melanin synthesis pathway and affect the virulence of pathogenic fungi [18–24]. Knockout of the *Cmr1* homolog or downstream regulated genes typically leads to a reduction or loss of melanin production. In the study of melanin production in *A. pullulans*, Jiang et al. initially identified the involvement of the *Cmr1* gene in melanin synthesis but did not conduct further in-depth research on its function [25].

This paper first isolated and characterized the melanin produced by *A. pullulans* Hit-lcy3$^T$, and then identified *Cmr1* involved in the synthesis of the melanin through bioinformatics analysis of its whole genome. Subsequently, knockout strains of *Cmr1* were obtained using homologous recombination, and the functions of *Cmr1* were investigated by comparing the morphology, biomass, melanin production, and expression levels of genes involved in melanin biosynthesis between the wild-type (WT), *Cmr1* knockout mutant(Δ*ApCmr1*), and *Cmr1* overexpression strains (*OEX-ApCmr1*). Finally, the scavenging activity of the melanin against free radicals including DPPH·, ABTS·, OH·, and $O_2^{-}$· was studied.

## 2. Materials and Methods

### 2.1. Stain, Plasmid, and Culture Medium

*A. pullulans* Hit-lcy3$^T$ and DH5α, as well as the PMD-18T vectors and pBARGPE1-EGFP, were obtained from our laboratory. The culture media used in this study included LB (10.0 g tryptone, 10.0 g NaCl, and 5.0 g yeast extract per liter, pH 7.2), YPD (20.0 g/L glucose, 10.0 g/L yeast extract, and 20.0 g/L peptone), seed medium (50.0 g/L sucrose, 2.0 g/L yeast extract, 1.0 g/L NaCl, 6.0 g/L $K_2HPO_4$, 0.2 g/L $MgSO_4$, and 0.6 g/L $(NH_4)_2SO_4$), and fermentation medium (100.0 g/L sucrose, 2.0 g/L yeast extract, 1.0 g/L NaCl, 6.0 g/L $K_2HPO_4$, 0.2 g/L $MgSO_4$, and 0.6 g/L $(NH_4)_2SO_4$).

### 2.2. Acquisition and Bioinformatics Analysis of Cmr1 Gene

The *Cmr1* gene was found in the genome of *A. pullulans* Hit-lcy3$^T$, and the primers Cmr1F/Cmr1R were designed to amplify the full-length gene. The amplified gene was sequenced and confirmed by comparison with NCBI (https://www.ncbi.nlm.nih.gov/, accessed on 10 December 2021). SMART (https://smart.embl-heidelberg.de/smart/set_mode.cgi?NORMAL=1, accessed on 5 March 2022) was used to predict the conserved domains of *Cmr1*. BDPG (https://www.fruitfly.org/seq_tools/promoter.html, accessed on 7 March 2022) was used to predict the promoter transcription start site. We then based on Plant CARE (http://bioinformatics.psb.ugent.be/webtools/plantcare/html/, accessed on 25 March 2022) to predict the promoter region of *Cmr1*. ClustalW (https://www.genome.jp/tools-bin/clustalw, accessed on 25 April 2022) and MEGA X software were used to perform multiple strain alignment and construct a phylogenetic tree of the amino acid sequence of *Cmr1*.

### 2.3. Melanin Extraction Method

*A. pullulans* Hit-lcy3$^T$ was cultured in a fermentation medium for 6, 7, 8, 9, 10, and 11 days, and the biomass was determined. The melanin was then extracted and purified using an acid–ethanol precipitation method. The fermentation broth was thoroughly mixed, and 50 mL of the mixture was centrifuged at 4000 rpm for 10 min. The supernatant was collected for further processing, and the precipitate was the biomass. The collected supernatant was then mixed with 1 mol/L hydrochloric acid to adjust the pH to 2–3 and shaken vigorously. After that, it was left to stand at 60 °C to induce complete precipitation. The product was then filtered through a 600-mesh filter paper and dried at 60 °C to constant weight to obtain the crude melanin. Finally, the crude product was repeatedly dissolved in a strong base, acid–ethanol precipitated and purified to obtain melanin with higher purity.

### 2.4. FT-IR Analysis of Melanin

The preparation method of the FT-IR sample was as follows: the isolated melanin was freeze-dried, ground into powder, and then a suitable amount of the powder was placed in a PerkinElmer 100 FT-IR spectrometer (PerkinElmer Inc., Waltham, MA, USA) for analysis under conditions of a resolution of 4 $cm^{-1}$ and a scanning range of 500–4000 $cm^{-1}$, to obtain the infrared spectrum of the melanin [26].

### 2.5. UV–Vis Analysis of Melanin

A suitable amount of melanin extract and standard substance were dissolved in 1 mol/L sodium hydroxide solution and then placed in a UV-6100 type UV–Vis spectrophotometer to scan the absorption spectrum in the wavelength range of 190–800 nm in spectral scan mode [27].

### 2.6. Plasmid Construction

The primers are listed in Table S1. To construct the *Cmr1* transcription factor knockout plasmid, the PMD-18T plasmid was used as the backbone, and the *Cmr1* gene was replaced with the hygromycin gene using homologous recombination to achieve the deletion of the *Cmr1* gene. First, the Hyg fragment and the upstream and downstream fragments of the *Cmr1* gene (Cmr1-up and Cmr1-down) were amplified by PCR. Then, the PMD-18T plasmid was double-digested with *EcoRI* and *BamHI*, and the Cmr1-up, Hyg, and Cmr1-down fragments were connected to the PMD-18T knockout plasmid backbone using the ClonExpress II one-step cloning kit, thus resulting in the knockout plasmid PMD-Hyg-$\triangle$Cmr1.

For the construction of the pBARGPE-EGFP-Cmr1 vector, first, a 3013 bp fragment covering the coding sequence of *Cmr*1 was amplified from the genome of *A. pullulans* Hit-lcy3$^{T}$. Then, *EcoRI* and *KpnI* were used to digest the *Cmr1* fragment and pBARGPE1-EGFP vector, respectively. Finally, ClonExpress II one-step cloning was used to ligate the recovered *Cmr1* fragment into pBARGPE1-EGFP, resulting in the overexpression of plasmid pBARGPE-EGFP-Cmr1.

### 2.7. Transformation and Validation of Mutant Strains

The successfully constructed overexpression vector pBARGPE-EGFP-Cmr1 and knockout vector PMD-Hyg-$\triangle$Cmr1 were transformed into *A. pullulans* protoplasts using the PEG-mediated protoplast transformation method [28]. The genomic DNA of transformants was extracted, and EGFP (egfpF/egfpR) and hygromycin resistance gene (hygR/hygF) were amplified using specific primers and validated by sequencing.

### 2.8. Effect of Cmr1 on Biomass and Melanin Production

Wild-type, *Cmr1* deletion strain, *ΔApCmr1*, and *Cmr1* overexpression strain, *OEX-ApCmr1*, were cultured in a fermentation medium for 9 days to determine their biomass and melanin production.

### 2.9. Morphological Observation

The morphological features of *A. pullulans* Hit-lcy3$^{T}$ and its mutants after 9 days of cultivation on YPD medium were observed using a light microscope (Nikon ECLIPSE E200, Nikon, Tokyo, Japan) and an electron microscope (JSM-6700F, JEOL, Tokyo, Japan). For scanning electron microscopy (SEM) sample preparation, square pieces of 0.5 cm $\times$ 0.5 cm were cut from the agar plate, fixed in 2.5% glutaraldehyde at 4 °C for about 12 h, washed twice with phosphate buffer, and treated with graded ethanol (50%, 70%, 90%, 100%). Finally, the samples were placed on conductive tape after treatment with acetone and freeze-dried, and observed under an SEM.

### 2.10. Analysis of the Transcriptional Levels of Genes Related to the ApCmr1 Biosynthetic Gene Cluster

Total RNA was extracted from the WT and different transformants grown in the fermentation medium. EasyScript First-Strand cDNA Synthesis SuperMix (TransGen Biotech, Beijing, China) was used for reverse transcription to obtain cDNA. RT-qPCR was performed using Go Taq qPCR Master Mix (Promega, Madison, Wisconsin, USA) in a real-time PCR system (Thermo Fisher Scientific, Wilmington, Massachusetts, USA), with the β-actin gene as the internal control gene. The relative expression levels of *Cmr1*, *PKS*, *SCD1*, and *THR1* were quantitatively detected, and expression levels were evaluated using the $2^{-\Delta\Delta CT}$ method.

### 2.11. Investigation of Antioxidant Activity of Melanin In Vitro

#### 2.11.1. DPPH· Scavenging Assay

A 0.2 mmol/L solution of DPPH• in absolute ethanol was prepared. Different concentrations of melanin samples and Vc (Vitamin C) solution were mixed with 100 μL of DPPH• solution in absolute ethanol, respectively. The mixtures were thoroughly vortexed, and shielded from light. After 30 min. the $OD_{517}$ was measured using a multimode plate reader [12].

$$\text{DPPH· scavenging rate (\%)} = [1 - (A_1 - A_2)/A_0] \times 100\%.$$

$A_0$: 100 μL anhydrous ethanol + 100 μL DPPH• absolute ethanol solution;

$A_1$: 100 μL test solution + 100 μL DPPH· absolute ethanol solution;
$A_2$: 100 μL test solution + 100 μL absolute ethanol solution.

2.11.2. ABTS· Scavenging Assay

First, 4 mmol/L ABTS· reserve solution and 2.6 mmol/L $K_2S_2O_8$ solution were prepared. Then, the ABTS· reserve solution and the $K_2S_2O_8$ solution were mixed in a 1:1 ratio and allowed to stand for about 12 h. Finally, it was diluted with distilled water to obtain ABTS· working solution with an absorbance of 0.700 at 734 nm wavelength. Next, 100 μL of different concentrations of melanin solution and Vc solution were mixed into 96-well plates, and then 100 μL ABTS· working solution was added to each well. After 10 min, the absorbance of the sample at 734 nm wavelength was measured [12].

$$\text{ABTS· scavenging rate (\%)} = [1 - (A_1 - A_2)/A_0] \times 100\%$$

$A_0$: 100 μL distilled water + 100 μL ABTS·;
$A_1$: 100 μL test solution + 100 μL ABTS·;
$A_2$: 100 μL test solution + 100 μL distilled water.

2.11.3. OH· Scavenging Assay

Then, 1 mL melanin solution and Vc solution of different concentrations were added into 1 mL 9 mmol/L $FeSO_4$ and 1 mL $H_2O_2$ in turn. The solution was mixed well and incubated at 37 °C for 10 min. Then, 1 mL 9 mmol/L salicylic acid was added and the reaction continued at 37 °C for 30 min. The absorbance of the sample at 536 nm wavelength was measured [12].

$$\text{OH· scavenging rate (\%)} = [1 - (A_1 - A_2)/A_0] \times 100\%.$$

$A_0$: 1 mL distilled water + 1 mL $FeSO_4$ + 1 mL $H_2O_2$ + 1 mL salicylic acid;
$A_1$: 1 mL of the test solution + 1 mL $FeSO_4$ + 1 mL $H_2O_2$ + 1 mL salicylic acid;
$A_2$: 1 mL distilled water + 1 mL $FeSO_4$ + 1 mL distilled water + 1 mL salicylic acid.

2.11.4. $O_2^-$· Scavenging Assay

In each 6.5 mL reaction mixture, 2 mL L-methionine (39 mmol/L), 2 mL NBT (225 μmol/L), 1 mL disodium EDTA (0.6 mmol/L), 1 mL riboflavin (12 μmol/L), and 0.5 mL of the melanin solution or Vc solution of varying concentrations were added. All solutions were prepared in 0.05 mol/L phosphate buffer (pH 7.8). The reaction mixture containing the sample was incubated at 25 °C for 15 min, and the absorbance was measured at 560 nm. The reaction mixture without the sample served as the control and was kept in the dark. The scavenging activity was calculated as follows [27]:

$$O_2^-\text{· scavenging rate (\%)} = (1 - A_1/A_0) \times 100\%$$

$A_0$: 0.5 mL distilled water + 2 mL L-methionine + 2 mL NBT + 1 mL EDTA + 1 mL riboflavin;

$A_1$: 0.5 mL sample solution + 2 mL L-methionine + 2 mL NBT + 1 mL EDTA + 1 mL riboflavin.

### *2.12. Statistical Analysis*

All data were statistically analyzed using SPSS 19.0 (IBM Corporation). Gene expression analysis was evaluated using one-way analysis of variance (ANOVA). Duncan's multiple-range test was used to compare data means. *p* value less than 0.05 was considered statistically significant.

## 3. Results and Discussion

### 3.1. Bioinformatics Analysis of Cmr1

The full length of the *Cmr1* gene in *A. pullulans* is 3013 bp, encoding a protein of 945 amino acids. There are two introns at positions 71 bp–122 bp (50 bp) and 986 bp–1015 bp (30 bp), respectively. SMART analysis of the protein sequence revealed that there are two adjacent $Cys_2His_2$ zinc finger domains located at positions 2–24 bp and 30–52 bp of the amino acid sequence, respectively, and a $Zn(II)_2Cys_6$ binuclear cluster domain located at positions 333–361 bp of the N-terminus of the *Cmr1* protein. The upstream 1500 bp sequence of *Cmr1* was analyzed using the BDPG tool, revealing that *Cmr1* is a member of the GATA family of transcription factors, and its gene promoter region contains three GATA motifs. The cis-acting element analysis of the promoter enhancer region using Plant CARE showed that there are one core promoter element (TATA-box) and three common cis-acting elements (CAAT-box) (Figure S1).

In the amino acid sequence's alignment with 20 different fungi, a high similarity was observed in the zinc finger structural sequence of the *Cmr1* gene (Figure 1A). However, *A. pullulans* Hit-lcy3$^T$ showed a difference at the 9th amino acid position compared to other strains, with histidine instead of glutamine, which is found in other fungi. Due to its distant evolutionary relationship with *Candida albicans* Ca529L and *Saccharomyces cerevisiae* 218 S288C, it was not included in the phylogenetic tree (Figure 1B).

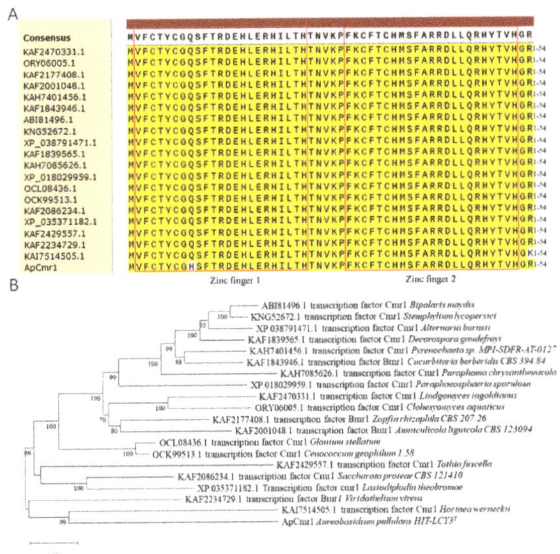

**Figure 1.** Bioinformatics analysis of *ApCmr1*. (**A**) Comparison of the zinc finger domain of *ApCmr1* and the homologous proteins in 20 fungi. (**B**) Phylogenetic tree of *ApCmr1* and its homologous proteins in 20 fungi.

### 3.2. Fermentation Time of A. pullulans Hit-lcy3T for Melanin Production

The melanin and biomass were measured after different days of fermentation. As shown in Figure 2, both the melanin and biomass increased and then decreased with the increase in fermentation time. At 9 days, the highest melanin yield and biomass were reached at 9.357 g/L and 16.605 g/L, respectively, which were significantly higher than those obtained from *Ophiocordyceps sinensis* and *Streptomyces* sp. (1.775 g/L and 1.46 g/L, respectively) [27,29].

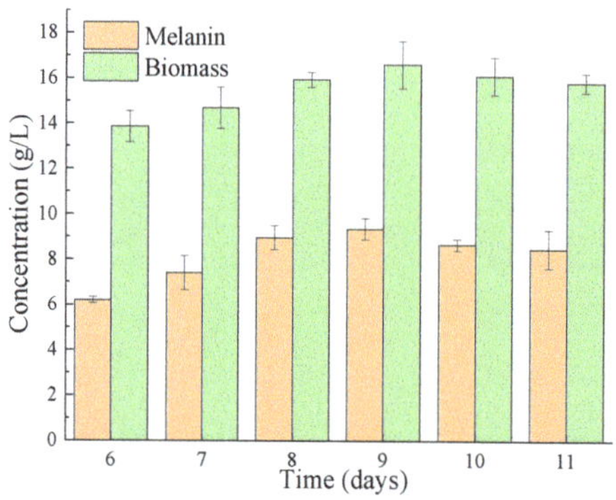


**Figure 2.** The effect of fermentation time on the biomass and melanin production of *A. pullulans* Hit-lcy3$^T$.

### 3.3. FT-IR Spectroscopy of DHN Melanin

The obtained melanin was characterized using FT-IR spectroscopy. As shown in Figure 3, the infrared spectrum of the DHN melanin extracted from *A. pullulans* Hit-lcy3$^T$ showed similar characteristic peaks as the melanin extracted from *A. pullulans* NBRC 100716, as reported by Mujdeci. The peak at 3300 cm$^{-1}$ is attributed to the stretching vibrations of –OH in aromatic ethers. Two obvious peaks at 2978 cm$^{-1}$ and 2900 cm$^{-1}$ indicate the presence of C–H stretching vibrations in aromatic hydrocarbons. In comparison with the melanin studied by Centeno and Pacelli, which exhibited two obvious peaks at 2950–2850 cm$^{-1}$, the two peaks observed in the present study were attributed to stretching vibrations of C–H in aromatic rings caused by lipids or amino acid residues during the extraction process [26,30]. The characteristic peaks of the melanin and synthetic melanin were the same at 1681 cm$^{-1}$, 1607 cm$^{-1}$, 1521 cm$^{-1}$, 1256 cm$^{-1}$, and 893 cm$^{-1}$. The strong band at 1681 cm$^{-1}$ is attributed to the vibration of C=O [27], while the strong band at 1607 cm$^{-1}$ is attributed to the vibration of the C=C group in aromatic rings [28]. The band at 1521 cm$^{-1}$ is attributed to the N–H bending vibration in the amide group of the aromatic rings and the peak at 1398 cm$^{-1}$ is attributed to the bending vibration of C–H in the aromatic hydrocarbons. The peak at 1256 cm$^{-1}$ is attributed to the stretching vibration of C–N in the amide group of the molecule. The peak observed at 987 cm$^{-1}$ is related to the out-of-plane bending vibration of C–H in the aromatic amine [31].

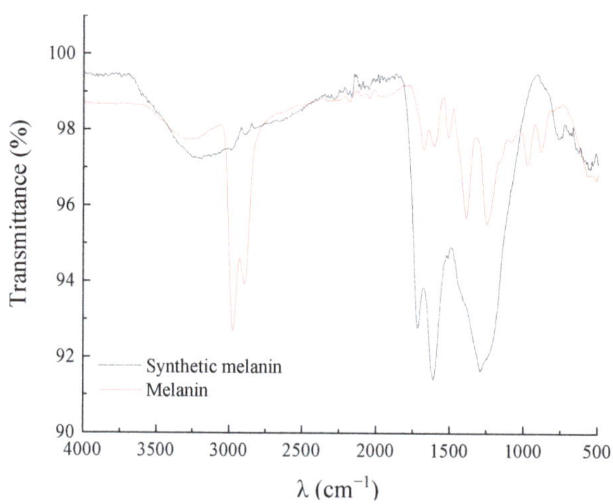


**Figure 3.** FT-IR of DHN melanin and synthetic melanin.

### 3.4. Analysis of UV–Vis Absorption Spectra of DHN Melanin

It has been observed in previous studies that the longer the wavelength, the lower the absorbance of melanin, which is a characteristic feature of melanin. Additionally, the absorption curve of melanin is similar to that of L-DOPA melanin nanoparticles. However, Li et al. found that the absorbance of purified *Streptomyces* sp. melanin showed almost linear behavior and did not exhibit any maximum or minimum absorption within the range of 200–800 nm [27]. Pacelli et al. compared the absorbance spectra of *Cryomyces antarcticus* melanin, L-DOPA, and DHN, and found that the highest peak of absorption was in the range of 200–300 nm [30]. Analysis of the UV–Vis absorption spectra of DHN melanin extracted from *A. pullulans* showed that the maximum absorbance of melanin occurred at 218 nm, which is consistent with the maximum absorbance of synthetic melanin (Figure 4). This is consistent with the results of Thokur's study on the DHN melanin of *Phyllosticta capitalensis* [32].

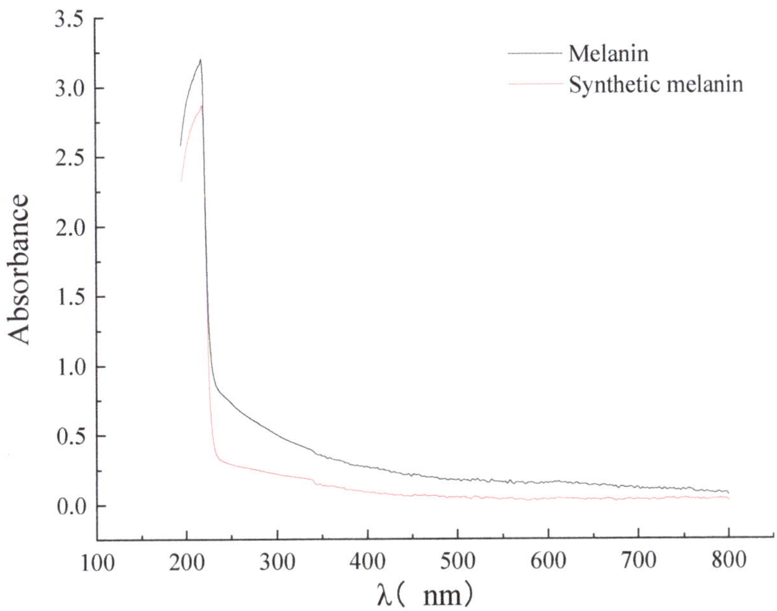


**Figure 4.** The UV–Vis of DHN melanin extracted from *A. pullulans* and the synthetic melanin.

### 3.5. Schematic and Verification of Gene Knockout

To verify the gene function of *Cmr1* in *A. pullulans* Hit-lcy3$^{T}$, a knockout vector was constructed using PMD-18T as the vector backbone through homologous recombination, as shown in Figure 5A. Cmr1-up, hyg, Cmr1-down, *HindIII*, and *EcoRI* double enzyme digestion PMD-18T vector bands were validated by agarose gel electrophoresis, with band sizes of 1305 bp, 1412 bp, 1289 bp, and 2635 bp, respectively (Figure 5B). As the △Cmr1 vector contains two *EcoRI* restriction sites while PMD-18T contains only one, *EcoRI* digestion was performed on both PMD-18T and △Cmr1 vectors separately. The resulting bands for △Cmr1 were 2284 bp and 4610 bp, while PMD-18T only produced a band of 2693 bp (Figure 5C). The amplification with the hygR/hygF primers revealed a 1000 bp hygromycin fragment in Δ*APCmr1*, which was not present in the WT strain (Figure 5D).

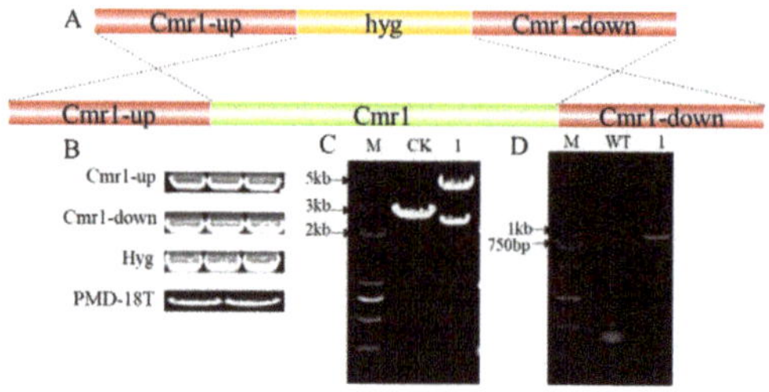

**Figure 5.** Construction and validation of the *Cmr1* knockout plasmid. (**A**) Construction strategy of the *Cmr1* knockout vector; (**B**) Cmr1-up, Cmr1-down, hyg, and PMD-18T bands after double enzyme digestion; (**C**) enzyme digestion (*EcoRI*) validation of the △Cmr1 vector, where CK represents the PMD-18T vector backbone and 1 represents the △Cmr1 vector; (**D**) PCR validation of the knockout strain, where WT represents the wild-type strain and 1 represents the Δ*APCmr1* knockout strain.

### 3.6. Construction and Verification of Cmr1 Overexpression Plasmid

Due to the presence of a strong gpdA promoter in pBARGPE-EGFP, it was selected as the backbone for the overexpression vector. The construction process of pBARGPE-EGFP-Cmr1 overexpression plasmid involved double digestion with *EcoRI* and *KpnI*, followed by ligation with the amplified *Cmr1* gene from the genome, as illustrated in Figure 6A. With the pBARGPE-EGFP and pBARGPE-EGFP-Cmr1 as templates, the amplification, using primers epc1-F and epc1-R, showed that a band of 1700 bp was obtained from pBARGPE-EGFP-Cmr1, whereas no band was detected from pBARGPE-EGFP (Figure 6B). Since pBARGPE-EGFP-Cmr1 contains two *BamHI* restriction sites, while pBARGPE-EGFP only has one *BamHI* restriction site, *BamHI* digestion was performed to verify both pBARGPE-EGFP and pBARGPE-EGFP-Cmr1. The results showed that pBARGPE-EGFP-Cmr1 produced two bands of 2672 bp and 6999 bp, respectively, after digestion, while pBARGPE-EGFP produced only a single band of 6700 bp. Furthermore, amplification of the green fluorescent protein-encoding gene using egfpF/egfpR primers revealed a band of 720 bp corresponding to the EGFP gene in the *OEX-ApCmr1* strain but not in the WT strain (Figure 6C). Fluorescence inverted microscope analysis (Figure 6E) confirmed the successful expression of the green fluorescent protein in the *OEX-ApCmr1* strain.

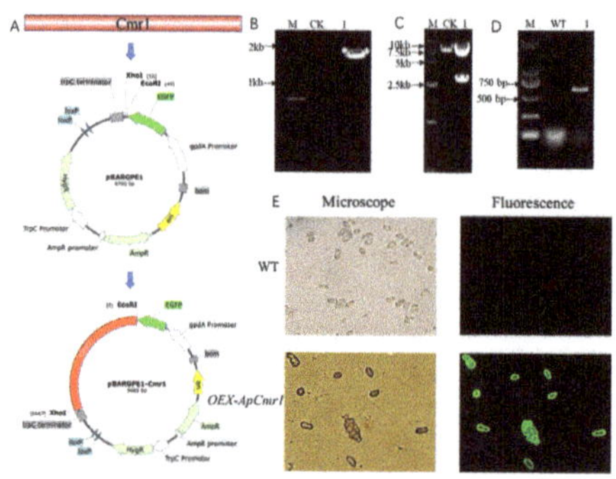

**Figure 6.** Constructing the *Cmr1* overexpression plasmid and validating the mutant strain. (**A**) Construction strategy for *Cmr1* overexpression vector; (**B**) PCR validation of pBARGPE-EGFP-Cmr1 plasmid, with CK representing the empty vector as negative control and 1 representing pBARGPE-EGFP-Cmr1 plasmid; (**C**) enzyme digestion validation (*BamHI*) of pBARGPE-EGFP vector, with CK representing the empty vector as negative control and 1 representing pBARGPE-EGFP-Cmr1 plasmid; (**D**) PCR validation of the overexpression strain, with WT representing wild-type strain, and 1 representing the *OEX-ApCmr1* overexpression strain; (**E**) green fluorescence validation of the overexpression strain, with WT representing wild-type strain and *OEX-ApCmr1* as the overexpression strain.

### 3.7. Effect of Cmr1 Mutant Strain on Morphology and Melanin Production

As shown in Figure 7, after 9 days of cultivation in a YPD medium, the surface of the Δ*ApCmr1* mutant strain was overall white, smooth, and moist, without aerial mycelia, while both the WT and *OEX-ApCmr1* strains had abundant aerial mycelia on the YPD medium plate. The *OEX-ApCmr1* mutant strain exhibited significantly higher DHN melanin secretion than the WT strain. Under the microscope, there were significant differences in melanin secretion among the two mutant strains and the WT strain, with no melanin generation in the Δ*ApCmr1* strain, while the *OEX-ApCmr1* strain had extremely high melanin content, staining the entire cell black. SEM observation revealed that the morphology of *A. pullulans* after knocking out the *ApCmr1* gene was mostly yeast-like, and the yeast cells of the mutant strain had obvious defects in their edges. Overexpression of *ApCmr1* led to chlamydospores becoming the main form, which was consistent with Mujdeci finding that the main form of melanin synthesis is chlamydospores [15]. It is speculated that the GATA-type transcription factor *Cmr1*, as a key transcription activation factor, not only regulates the synthesis of melanin but may also have an effect on the cell morphology of *A. pullulans*.

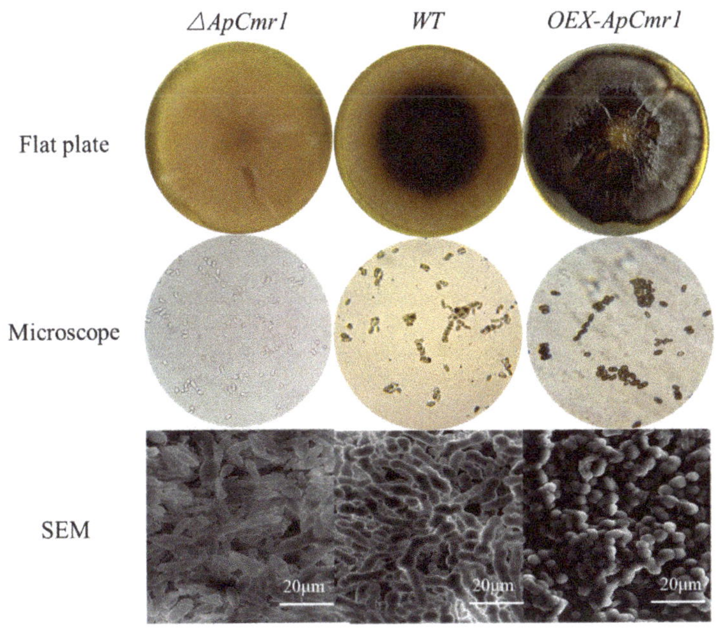


**Figure 7.** Morphological observation of the mutant strain.

WT and each mutant strain were cultured in a fermentation medium for 9 days, and the production of melanin and biomass is shown in Figure 8. The melanin of the Δ*APCmr1* strain disappeared, and its biomass was reduced compared to the WT and *OEX-APCmr1*. This is partly due to the decrease in the expression level of melanin-related genes and the possible decrease in biomass due to the developmental defects of the strain caused by the knockout of the *Cmr1* gene. However, the melanin production in the *OEX-ApCmr1* strain was 1.436 times that of the WT, and its biomass also slightly increased, which was one of the reasons for the increase in melanin production.

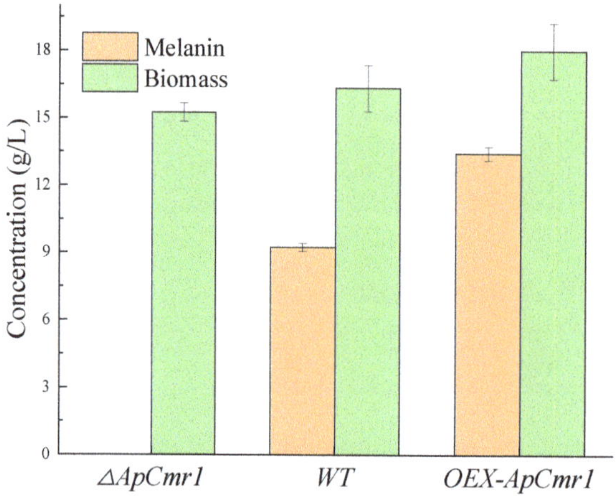


**Figure 8.** Biomass and melanin production of mutant strains.

*3.8. Expression of Melanin Synthesis-Related Gene*

After 9 days of fermentation in the culture medium, RNA was extracted from WT and various mutant strains. Reverse transcription was performed to obtain cDNA, and RT-qPCR analysis was conducted on WT, Δ*ApCmr1*, and *OEX-ApCmr1* strains (Figure 9). The results revealed that the expression levels of *Cmr1*, *PKS*, *SCD1*, and *THR1* genes were nearly undetectable after knocking out the *Cmr1* gene, providing further evidence that *Cmr1* acts as a transcriptional regulator of melanin synthesis in *A. pullulans*, which is consistent with the findings of Chi's study [33]. Overexpression of the *Cmr1* gene resulted in a 5.488-fold increase in the relative expression of *Cmr1*, and 7.324-fold, 8.535-fold, and 8.434-fold increases in the relative expression of *PKS*, *SCD1*, and *THR1* genes, respectively. However, the expression levels were still lower than the relative expression levels of *Bmr1*, *PKS*, *SCD1*, and *THR1* genes in the *Bipolaris oryzae* strain after overexpression of the *Bmr1* gene [19].

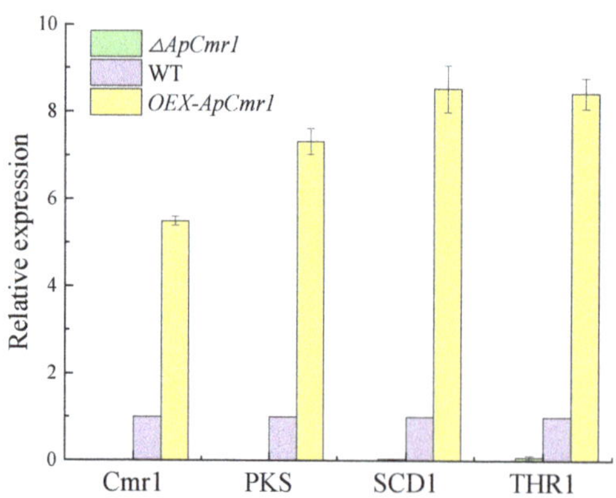


**Figure 9.** The effect of *Cmr1* on the expression of melanin synthesis genes.

*3.9. In Vitro Antioxidant Activity of Melanin*

Studies have shown that melanin from microorganisms has certain antioxidant activity. To explore the antioxidant activity of melanin, Vc was selected as a control to evaluate its ability to scavenge DPPH·, ABTS·, OH·, and $O_2^{-}$·. The results of melanin scavenging DPPH· are shown in Figure 10A. As the concentration of Vc and melanin increased, the scavenging rate of DPPH· also increased, reaching a maximum of 84.96%, which is slightly lower than the ability of melanin extracted from *Cryptococcus rajasthanensis* KY627764 to scavenge DPPH· [34]. The melanin scavenging activity against ABTS· is shown in Figure 10B. With the increase in melanin concentration, the scavenging rate of ABTS· also increased, reaching a maximum of 80.44%, but it was still lower than that of Vc. The melanin scavenging activity against OH· is shown in Figure 10C. With the increase in melanin concentration, the scavenging rate of OH· also increased, reaching a maximum of 81.57%. The ability of melanin to scavenge $O_2^-$· increased with the concentration (Figure 10D), reaching a scavenging rate of 17.83% at 100 μg/mL, but was weaker compared to the ability of melanin extracted from *Streptomyces* sp. by Li et al. to scavenge $O_2^-$· [27]. These results suggest that the melanin produced by *A. pullulans* has a strong ability to scavenge DPPH·, ABTS·, and OH·, but a weaker ability to scavenge $O_2^-$·.

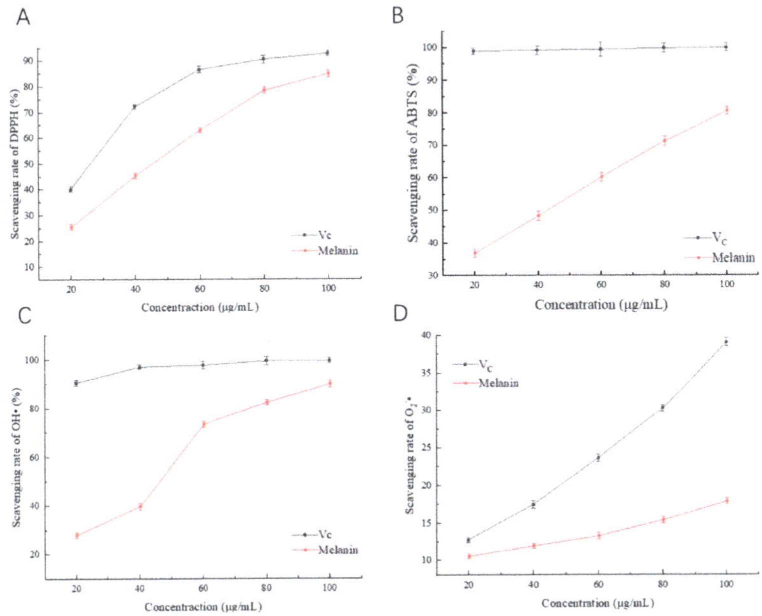


**Figure 10.** Antioxidant activity of melanin. (**A**) Scavenging rate of DPPH·. (**B**) Scavenging rate of ABTS·. (**C**) Scavenging rate of OH·. (**D**) Scavenging rate of $O_2^-$·.

Currently, research on the synthesis and application of melanin is increasing, and the potential of melanin as a new type of functional active substance has been gradually explored. Most fungi synthesize melanin through two pathways: tyrosine (L-DOPA pathway) or malonyl CoA (DHN pathway) [35]. In the DOPA pathway, tyrosinase first catalyzes L-tyrosine (monophenol) and L-DOPA (Bisphenol) to generate L-dopaquinone under the catalysis of the enzyme. Then, L-dopaquinone undergoes enzyme-catalyzed and chemical reactions to form a complex chemical structure, a non-homogeneous polymeric material known as L-DOPA melanin. In the DHN pathway, the precursor of malonyl CoA is endogenously produced. Under the catalysis of polyketide synthase, five malonyl CoA are consecutively decarboxylated and condensed to generate 1,3,6,8-tetrahydroxynaphthalene (THN). THN then undergoes reduction and dehydration to generate 1,8- DHN, which is finally polymerized to obtain DHN melanin. *Cmr1* is a key transcriptional regulatory factor of DHN melanin that regulates the genes involved in the DHN melanin synthesis pathway in most fungi. It specifically binds to the upstream of the polyketide synthase gene to initiate the synthesis of melanin [36]. Studies have shown that in melanin within

*A. pullulans*, DHN melanin is the primary form, which is synthesized by the polymerization of 1,8-DHN, and is important for maintaining cell wall integrity and morphology.

Knocking out the *Cmr1* gene caused the disappearance of melanin in *A. pullulans*, as the expression levels of genes related to melanin synthesis were almost zero after knocking out *Cmr1*. Moreover, the biomass of the mutant strains decreased, and they mainly showed a yeast-like morphology. Since fungal melanin is a part of the cell wall composition, the strains that underwent knockout displayed developmental defects around the cell surface, which is consistent with Liang's study. The reason for this was the loss of melanin, which led to the loss of the originally filled areas of melanin in the cell wall, resulting in irregular developmental defects in the cell wall, demonstrating that melanin plays an important role in maintaining the fungal cell wall morphology [37]. The *OEX-ApCmr1* strain showed increased expression levels of *Cmr1*, *PKS*, *SCD1*, and *THR1* genes related to melanin synthesis by 5.488-fold, 7.324-fold, 8.535-fold, and 8.434-fold, respectively, and melanin production was 1.436-times higher than that of the WT strain. This may be due to the overexpression of melanin in the *OEX-ApCmr1* strain, which bound more to the cell wall and was not released into the fermentation broth. Morphological observations showed that the cells of the *OEX-ApCmr1* strain were predominantly chlamydospores and hyphae, consistent with previous research that suggested that chlamydospores and hyphae were the primary forms of melanin production [38].

The mechanism by which melanin acts as an antioxidant is still unclear. However, recent research has found that its antioxidant activity is related to the monomer 1,8-DHN. This is due to the unique pattern of hydroxylation of the hydrogen-bonding ring in 1,8-DHN, which facilitates the rapid transfer of H atoms [39]. Our research has found that melanin has good scavenging activity on OH·, DPPH·, and ABTS·, but a weak effect on $O_2^-$·. The reason for this is that when $O_2^-$· extracts H atoms to form OOH· or $H_2O_2$, the transfer of the two hydrogen atoms only occurs at the adjacent carbon position. However, during the polymerization of 1,8-DHN to form melanin, one or two of the adjacent hydroxy groups are replaced, resulting in a weaker scavenging activity on $O_2^-$· [40]. In the food industry, natural pigments are valued for their safety, nutrition, and certain bioactivity. Therefore, melanin can be used as a natural coloring agent and additive in food production. Melanin has already been used to prepare yogurt to enhance its antioxidant and free radical scavenging properties, as well as to develop various new types of food, such as black bean vermicelli, black tofu skin, and black caviar [7]. It can also be used as a functional food to better exert its role in reducing blood cholesterol, lowering blood sugar, and anti-tumor activity. There are many microorganisms in nature that can produce pigments. By using genetic engineering methods to obtain genes encoding pigment synthesis and constructing microorganism pigment production engineering bacteria, and effectively regulating their metabolic pathways, it is possible to achieve industrial production of pigments. With the rapid development of biotechnology, microbial pigments are expected to replace synthetic pigments as mainstream food coloring agents.

## 4. Conclusions

In this study, we investigated the regulatory gene *Cmr1* in melanin synthesis through the construction of gene knockout and overexpression strains. Our results showed that *Cmr1* is a key regulator of melanin synthesis in Hit-lcy3$^T$, and its absence caused developmental defects. Furthermore, the expression levels of *Cmr1*, *PKS*, *SCD1*, and *THR1* genes related to melanin synthesis were regulated by *Cmr1*. Finally, we found that melanin possesses strong scavenging activity against DPPH·, ABTS·, and OH·, but weaker activity against $O_2^-$·. These findings suggest that Hit-lcy3$^T$ melanin has the potential to be used as a functional food additive. This study provides a theoretical basis and guidance for the efficient production of melanin strains of *A. pullulans* and research on the antioxidant activity of melanin.

**Supplementary Materials:** The following supporting information can be downloaded at: https://www.mdpi.com/article/10.3390/foods12112135/s1, Figure S1: promoter prediction analysis of *ApCmr1*; Table S1: primers used in this study.

**Author Contributions:** Conceptualization, W.W. and K.Z.; methodology, C.L.; software, S.Z.; validation, J.G.; formal analysis, W.Z.; investigation, X.R.; resources, H.C.; supervision, Q.Y. All authors have read and agreed to the published version of the manuscript.

**Funding:** This research received no external funding.

**Data Availability Statement:** The data presented in this study are available on request from the corresponding author.

**Conflicts of Interest:** The authors declare no conflict of interest.

## References


1. Butler, M.J.; Day, A.W. Fungal melanins: A review. *Can. J. Microbiol.* **1998**, *44*, 1115–1136. [CrossRef]
2. Dadachova, E.; Casadevall, A. Ionizing radiation: How fungi cope, adapt, and exploit with the help of melanin. *Curr. Opin. Microbiol.* **2008**, *11*, 525–531. [CrossRef]
3. Eisenman, H.C.; Casadevall, A. Synthesis and assembly of fungal melanin. *Appl. Microbiol. Biotechnol.* **2012**, *93*, 931–940. [CrossRef] [PubMed]
4. Echavarria, A.P.; Pagan, J.; Ibarz, A. Melanoidins Formed by Maillard Reaction in Food and Their Biological Activity. *Food Eng. Rev.* **2012**, *4*, 203–223. [CrossRef]
5. Fogliano, V.; Morales, F.J. Estimation of dietary intake of melanoidins from coffee and bread. *Food Funct.* **2011**, *2*, 117–123. [CrossRef]
6. Huang, L.; Liu, M.Y.; Huang, H.Y.; Wen, Y.Q.; Zhang, X.Y.; Wei, Y. Recent Advances and Progress on Melanin-like Materials and Their Biomedical Applications. *Biomacromolecules* **2018**, *19*, 1858–1868. [CrossRef]
7. Tran-Ly, A.N.; Reyes, C.; Schwarze, F.W.M.R.; Ribera, J. Microbial production of melanin and its various applications. *World J. Microbiol. Biotechnol.* **2020**, *36*, 170. [CrossRef]
8. Lee, J.H.; Hyun, C.K. Insulin-Sensitizing and Beneficial Lipid-Metabolic Effects of the Water-Soluble Melanin Complex Extracted from Inonotus obliquus. *Phytother. Res.* **2014**, *28*, 1320–1328. [CrossRef]
9. Xu, C.; Chen, T.T.; Li, J.L.; Jin, M.Z.; Ye, M. The structural analysis and its hepatoprotective activity of melanin isolated from *Lachnum* sp. *Process. Biochem.* **2020**, *90*, 249–256. [CrossRef]
10. Mesias, M.; Delgado-Andrade, C. Melanoidins as a potential functional food ingredient. *Curr. Opin. Food Sci.* **2017**, *14*, 37–42. [CrossRef]
11. Dufosse, L. Microbial production of food grade pigments. *Food Technol. Biotechnol.* **2006**, *44*, 313–321.
12. Zhang, K.; Zhou, W.; Wang, W.; Zhao, S.S.; Lin, C.Y.; Ru, X.; Guan, J.Q.; Cong, H.; Yang, Q. Area Gene Regulates the Synthesis of beta-Glucan with Antioxidant Activity in the *Aureobasidium pullulans*. *Foods* **2023**, *12*, 660. [CrossRef] [PubMed]
13. Lin, C.; Zhang, K.; Zhao, S.; Wang, W.; Ru, X.; Song, J.; Cong, H.; Yang, Q. Screening and identification of a strain of *Aureobasidium pullulans* and its application in potato starch industrial waste. *Environ. Res.* **2022**, *214*, 113947. [CrossRef] [PubMed]
14. Mujdeci, G.N. Experimental modeling and optimization of melanin production by *Aureobasidium pullulans* NBRC 100716 in carrot peel extract. *Environ. Prog. Sustain. Energy* **2022**, *41*, e13919. [CrossRef]
15. Mujdeci, G.N. Natural Melanin Synthesized by *Aureobasidium pullulans* Using Food Wastes and its Characterization. *Appl. Food Biotechnol.* **2021**, *8*, 307–318.
16. Bialy, H.E.; Elsayed, M.; Khalifa, M.A. Isolation and Characterization of melanized yeast form of *Aureobasidium pullulans* and physiological studies on the melanization process. *J. Nucl. Technol. Appl. Sci.* **2017**, *5*, 57–72.
17. Tsuji, G.; Kenmochi, Y.; Takano, Y.; Sweigard, J.; Farrall, L.; Furusawa, I.; Horino, O.; Kubo, Y. Novel fungal transcriptional activators, Cmr1p of *Colletotrichum lagenarium* and $pig_1$p of *Magnaporthe grisea*, contain $Cys_2His_2$ zinc finger and $Zn(II)^2Cys_6$ binuclear cluster DNA-binding motifs and regulate transcription of melanin biosynthesis genes in a developmentally specific manner. *Mol. Microbiol.* **2000**, *38*, 940–954.
18. Eliahu, N.; Igbaria, A.; Rose, M.S.; Horwitz, B.A.; Lev, S. Melanin biosynthesis in the maize pathogen *Cochliobolus heterostrophus* depends on two mitogen-activated protein kinases, $Chk_1$ and $Mps_1$, and the transcription factor $Cmr_1$. *Eukaryot. Cell* **2007**, *6*, 421–429. [CrossRef]
19. Kihara, J.; Moriwaki, A.; Tanaka, N.; Tanaka, C.; Ueno, M.; Arase, S. Characterization of the BMR1 gene encoding a transcription factor for melanin biosynthesis genes in the phytopathogenic fungus *Bipolaris oryzae*. *FEMS Microbiol. Lett.* **2008**, *281*, 221–227. [CrossRef]
20. Cho, Y. The transcription factor Amr1 induces melanin biosynthesis and conidium production but differentially suppresses virulence in *Alternaria brassicicola*. *Phytopathology* **2011**, *101*, S36.
21. Valiante, V.; Baldin, C.; Hortschansky, P.; Jain, R.; Thywissen, A.; Strassburger, M.; Shelest, E.; Heinekamp, T.; Brakhage, A.A. The *Aspergillus fumigatus* conidial melanin production is regulated by the bifunctional bHLH DevR and MADS-box RlmA transcription factors. *Mol. Microbiol.* **2016**, *102*, 321–335. [CrossRef] [PubMed]

22. Kurahashi, A.; Shimoda, T.; Sato, M.; Fujimori, F.; Hirama, J.; Nishibori, K. A putative transcription factor Gf.BMR1 in *Grifola frondosa*, the homolog of BMR1 in *Bipolaris oryzae*, was strongly induced by near-ultraviolet light and blue light. *Mycoscience* **2015**, *56*, 177–182. [CrossRef]
23. Schumacher, J. DHN melanin biosynthesis in the plant pathogenic fungus Botrytis cinerea is based on two developmentally regulated key enzyme (PKS)-encoding genes. *Mol. Microbiol.* **2016**, *99*, 729–748. [CrossRef] [PubMed]
24. Wang, Y.L.; Hu, X.P.; Fang, Y.L.; Anchieta, A.; Goldman, P.H.; Hernandez, G.; Klosterman, S.J. Transcription factor $VdCmr_1$ is required for pigment production, protection from UV irradiation, and regulates expression of melanin biosynthetic genes in Verticillium dahliae. *Microbiology* **2018**, *164*, 863–864. [CrossRef] [PubMed]
25. Jiang, H.; Chi, Z.; Liu, G.L.; Hu, Z.; Zhao, S.Z.; Chi, Z.M. Melanin biosynthesis in the desert-derived Aureobasidium melanogenum XJ5-1 is controlled mainly by the CWI signal pathway via a transcriptional activator $Cmr_1$. *Curr. Genet.* **2020**, *66*, 173–185. [CrossRef]
26. Centeno, S.A.; Shamir, J. Surface enhanced Raman scattering (SERS) and FTIR characterization of the sepia melanin pigment used in works of art. *J. Mol. Struct.* **2008**, *873*, 149–159. [CrossRef]
27. Li, C.; Ji, C.; Tang, B. Purification, characterisation and biological activity of melanin from *Streptomyces* sp. *FEMS Microbiol. Lett.* **2018**, *365*, fny077. [CrossRef]
28. Zhang, K.; Lin, C.Y.; Zhao, S.S.; Wang, W.; Zhou, W.; Ru, X.; Cong, H.; Yang, Q. The role of pH transcription factor Appacc in upregulation of pullulan biosynthesis in *Aureobasidium pullulans* using potato waste as a substrate. *Int. J. Biol. Macromol.* **2023**, *242*, 124794. [CrossRef]
29. Dong, C.H.; Yao, Y.J. Isolation, characterization of melanin derived from Ophiocordyceps sinensis, an entomogenous fungus endemic to the Tibetan Plateau. *J. Biosci. Bioeng.* **2012**, *113*, 474–479. [CrossRef]
30. Pacelli, C.; Cassaro, A.; Maturilli, A.; Timperio, A.M.; Gevi, F.; Cavalazzi, B.; Stefan, M.; Ghica, D.; Onofri, S. Multidisciplinary characterization of melanin pigments from the black fungus *Cryomyces antarcticus*. *Appl. Microbiol. Biotechnol.* **2020**, *104*, 6385–6395. [CrossRef]
31. Pal, A.K.; Gajjar, D.U.; Vasavada, A.R. DOPA and DHN pathway orchestrate melanin synthesis in Aspergillus species. *Med. Mycol.* **2014**, *52*, 10–18. [PubMed]
32. Suryanarayanan, T.S.; Ravishankar, J.P.; Venkatesan, G.; Murali, T.S. Characterization of the melanin pigment of a cosmopolitan fungal endophyte. *Mycol. Res.* **2004**, *108*, 974–978. [CrossRef] [PubMed]
33. Chi, Z.; Kong, C.C.; Wang, Z.Z.; Wang, Z.; Liu, G.L.; Hu, Z.; Chi, Z.M. The signaling pathways involved in metabolic regulation and stress responses of the yeast-like fungi *Aureobasidium* spp. *Biotechnol. Adv.* **2022**, *55*, 107898. [CrossRef] [PubMed]
34. Barretto, D.A.; Vootla, S.K. Biological activities of melanin pigment extracted from Bombyx mori gut-associated yeast Cryptococcus rajasthanensis KY627764. *World J. Microbiol. Biotechnol.* **2020**, *36*, 159. [CrossRef]
35. El-Bialy, H.A.; El-Gamal, M.S.; Elsayed, M.A.; Saudi, H.A.; Khalifa, M.A. Microbial melanin physiology under stress conditions and gamma radiation protection studies. *Radiat. Phys. Chem.* **2019**, *162*, 178–186. [CrossRef]
36. Jia, S.L.; Chi, Z.; Chen, L.; Liu, G.L.; Hu, Z.; Chi, Z.M. Molecular evolution and regulation of DHN melanin-related gene clusters are closely related to adaptation of different melanin-producing fungi. *Genomics* **2021**, *113*, 1962–1975. [CrossRef]
37. Liang, X.; Li, B.; Zhao, X.; Yao, L.; Kong, Y.; Liu, W.; Zhang, R.; Sun, G. 1,8-Dihydroxynaphthalene Melanin Biosynthesis in *Colletotrichum fructicola* Is Developmentally Regulated and Requires the Cooperative Function of Two Putative Zinc Finger Transcription Factors. *Phytopathology* **2022**, *112*, 2174–2186. [CrossRef]
38. Dehoog, G.S.; Yurlova, N.A. Conidiogenesis, Nutritional Physiology and Taxonomy of Aureobasidium and Hormonema. *Anton. Leeuw. Int. J. G.* **1994**, *65*, 41–54. [CrossRef]
39. Lino, V.; Manini, P.; Galeotti, M.; Salamone, M.; Bietti, M.; Crescenzi, O.; Napolitano, A.; d'Ischia, M. Antioxidant Activities of Hydroxylated Naphthalenes: The Role of Aryloxyl Radicals. *Chempluschem* **2023**, *88*, e202200449. [CrossRef]
40. Tiwari, M.K.; Jena, N.R.; Mishra, P.C. Mechanisms of scavenging superoxide, hydroxyl, nitrogen dioxide and methoxy radicals by allicin: Catalytic role of superoxide dismutase in scavenging superoxide radical. *J. Chem. Sci.* **2018**, *130*, 105. [CrossRef]

*foods* MDPI

*Article*

# Extrusion Modification of Wheat Bran and Its Effects on Structural and Rheological Properties of Wheat Flour Dough

Ranran Li [1,†], Chenyang Wang [1,†], Yan Wang [1], Xuan Xie [1], Wenjie Sui [1,*], Rui Liu [1], Tao Wu [1] and Min Zhang [2,*]

[1] State Key Laboratory of Food Nutrition and Safety, College of Food Science and Engineering, Tianjin University of Science & Technology, Tianjin 300457, China; li17862686224@126.com (R.L.); wcy2020d@163.com (C.W.); ywang@mail.tust.edu.cn (Y.W.); lr@tust.edu.cn (R.L.); wutaoxx@gmail.com (T.W.)

[2] College of Food Science and Bioengineering, Tianjin Agricultural University, Tianjin 300392, China

* Correspondence: wjsui@tust.edu.cn (W.S.); zm0102@sina.com (M.Z.)

† These authors contributed equally to this work.

**Abstract:** The study investigated the extrusion modification of wheat bran and its effects on structural and rheological properties of wheat flour dough. Extruded bran showed better solubility of dietary fiber and structural porosity, leading to higher hydration and swelling power. Addition of extruded bran to dough caused water redistribution as an intensive aggregation of bound water to gluten matrix with reduced mobility. The bran–gluten interaction influenced by water sequestering caused partial gluten dehydration and conversion of $\beta$-turn into $\beta$-sheet, which demonstrated the formation of a more polymerized and stable gluten network. Farinographic data confirmed the promotion of dough stability with extruded bran addition at lower gluten content, while viscoelastic data suggested improved dough elasticity at all gluten contents by increasing elastic moduli and decreasing loss tangent. This study would be useful for interpreting the modification effect and mechanism of extrusion on cereal brans and provide valuable guidance for applying it as an effective modification technology on the commercial production of cereal bran and its flour products.

**Keywords:** extrusion; wheat bran; dough; structure; rheological properties

## 1. Introduction

To satisfy the rising requirement for a healthy and low-calorie diet, the development of cereal products rich in dietary fiber has been an efficient means to improve consumers' dietary patterns [1–3]. As a good source of dietary fiber, wheat bran is the major by-product in flour mill processing. Wheat bran consists of epidermis, peel, seed, nucellus layer and aleurone layer [3,4]. It is principally formed of the higher contents of cellulose, vitamins and minerals, more so than endosperm [5,6]. Prior research indicated that dietary fibers in wheat bran had a positive impact on diabetes and could prevent colon cancer, and reduce diverticulosis and the formation of gallstones [6,7]. However, most of the wheat bran produced today is generally underutilized as animal feed or discarded as waste agricultural residues. Its use in human food is limited [5]. This phenomenon should be mainly owing to the high content of insoluble dietary fiber resulting in the deterioration of products' quality, which is correlated with their structural and rheological properties. Wheat bran could detrimentally affect the gluten network, reduce the dough resilience and weaken the framework of gas cells, and thus, gas retention. These effects inevitably lead to inferior baking quality, such as leavening restriction, loaf volume reduction, gritty texture and undesirable mouthfeel and taste [8]. Therefore, it is quite necessary to understand the basis for negative effects of dietary fibers on bread quality, so as to develop efficient strategies to eliminate quality defects, especially those correlated to dough rheology and bread volume [9–11].

Extrusion is a high-temperature/pressure short-time process for the modification and production of various kinds of food products. In the continuous screw shearing and heating stage, material is gradually exposed to a large pressure at the die head, which lasts for about 30~300 s and the heating temperature is in the range of 50~180 °C [12]. A number of chemical changes occur in this stage: starch gelatinization, protein cross-linking and flavor generation, etc. [9,13]. When the material is extruded out of the machine head, the instantaneous decompression causes water flashing and steam expansion, resulting in a porous structure formed of product. Extrusion has certain advantages, including the reduction of microbial content, inactivation of some enzymes and anti-nutritional factors and modification of sensory characteristics [1,2,9]. As a multi-stage, multi-functional and thermal/mechanical process, it has permitted a variety of food applications [12,14]. Previous studies showed that extrusion could change the physico-chemical properties of wheat bran and prompt the bioavailability and functionality of soluble dietary fiber (SDF) in bran materials [1,4,14]. However, the specific influencing mechanism of treated bran on the structure and rheology of wheat dough still remained to be further unveiled.

It is known that both hydration and polymerization of gluten protein are critical processes for dough formation [15–17]. During the hydration process, water acts as a solvent that solubilizes low molecular proteins and induces conformational changes in the hydrophobic interactions between bran and gluten protein. In the following, polymerization of gluten protein has an impact on the construction of tridimensional viscoelastic network, which would interfere with dough mobility [15,18], so the interaction between bran and gluten during hydration and polymerization are closely related to the extensibility and elasticity of dough and the textural features of terminal products. Until now, there are several hypotheses describing the impact of this interaction on dough quality, which was strongly associated with a combination of water binding and size depletion [19,20]. It was proposed that arabinoxylans in dietary fibers of bran tightly bound water in the dough so as to decrease the available water participating in the development of the gluten network [10,19,20]. The dilution of gluten, hindrance of the gluten network and piercing of gas cells by the addition of fiber particles contributed to the size depletion phenomena [10]. In addition, ferulic acids were supposed to narrow the development time, raise the breakage rate and stiffness of dough, with the interference mechanism tentatively attributed to its interference with the significant disulphide interchange reaction [3,10]. In the bakery industry, it is significant to understand the rheological changes in wheat dough during processing, owing to their close correlation with quality attributes of flour products. The reduction of dough elasticity by disturbing gluten network formation would lead to the interruption of gas cells, low volume of bread and poor quality of baking products [10,12,14]. Therefore, it is necessary to explore the interactions of extruded bran and gluten protein in the dough system and on these bases evaluate the influence of extruded bran on structural and rheological properties of wheat dough, for the aim of providing valuable guidance for applying extrusion as an effective modification technology on the commercial production of bran-containing flour products.

The novelty of this work is to reveal the interaction variations of extruded bran and gluten protein in dough systems with different gluten levels and further explore the effects of extruded bran on the structural and rheological properties of wheat dough. After extrusion modification, the composition, micro-structure and hydration characteristics of wheat bran were measured; then 0~25% extruded bran was added to dough with different gluten contents and the major changes of water states and gluten conformation of dough were analyzed; lastly, the farinographic and dynamic viscoelastic properties of dough were characterized, to reveal the possible physical mechanism of structure and rheology changes in dough system with extruded bran addition.

## 2. Materials and Methods

### 2.1. Materials

Wheat bran was purchased from Xiangyuan Jinxiangzhai Trading Co., Ltd., Shanxi Province, China. Raw material is in the form of thin slices with a diameter of 3~5 mm and a thickness of 0.5~1 mm. Different-gluten wheat flours were purchased from Binzhou

Zhongyu Food Co., Ltd., Shandong Province, China. The protein content, wet gluten content and alveograph parameters of wheat flour are listed in Table 1.

**Table 1.** The protein content, wet gluten content and alveograph parameters of wheat flours.

| Flour Types | Protein Content/% | Wet Gluten Content/% | Alveograph Parameters | | |
|---|---|---|---|---|---|
| | | | P/mm | L/mm | W/($10^{-4}$ × J) |
| Low-gluten wheat flour | 6.73 ± 0.23 | 21.56 ± 1.21 | 54.23 ± 2.36 | 95.63 ± 2.43 | 136.92 ± 2.37 |
| Middle-gluten wheat flour | 10.48 ± 0.56 | 26.89 ± 0.58 | 23.01 ± 0.89 | 54.99 ± 1.89 | 89.08 ± 2.02 |
| High-gluten wheat flour | 14.83 ± 0.45 | 34.56 ± 1.37 | 75.50 ± 2.75 | 160.88 ± 3.65 | 213.25 ± 4.03 |

### 2.2. Extrusion Modification

The extrusion of wheat bran was by a DS56-X twin-screw co-rotating, self-wiping extruder (Jinan Saixin Machinery Co., Jinan, China) with a ratio of length to diameter of 25 and screw speed up to 600 rpm. Then, 1 kg wheat bran was processed by the extruder each batch with a water content of 25 ($w/w$, %). The barrel temperature in I, II and III heating blocks was set to 100 °C, 145 °C and 170 °C, respectively, and the feeding speed and screw speed of 10 r/min and 100 r/min were selected based on preliminary study [21]. The extrudates were oven-dried at 50 °C overnight followed by being finely ground and passed through a 60-mesh screen (particle size < 250 μm).

### 2.3. Compositional Determination of Wheat Bran

The TDF, SDF and IDF contents were determined by the AOAC 991.43 method. Protein, starch and lipid content were measured according to AOAC 920.87, 996.11 and 983.23 methods.

### 2.4. Microscopic Structure Characterization of Wheat Bran

The SEM photographs of bran samples were taken using Hitachi SU-1510 system (Hitachi, Pleasanton, CA, USA).

### 2.5. Hydration Property Determination of Wheat Bran

Water-holding capacity (WHC), oil-holding capacity (OHC) and water swelling capacity (WSC) were determined according to Sui et al. [3].

### 2.6. Preparation of Wheat Dough with Addition of Wheat Bran

Dough was prepared by mixing different-gluten flour and wheat bran (0%, 5%, 10%, 15%, 20%, 25%) with water until the consistency of the dough peaked in a Brabender farinograph using a 50 g mixing bowl. At the end of mixing, doughs were collected and then sealed with preservative film for around 0.5 h to wake up. The characterization of LF-NMR was conducted within 30~45 min.

### 2.7. Fourier-Transformed Infrared Spectroscopy (FT-IR) of Wheat Dough

The protein secondary structure of dough samples was characterized by FTIR spectroscopy (IS50, Nicolet, Thermo Fisher, Waltham, MA, USA). For each measurement, the freeze-dried dough was mixed with KBr (1:150, $w/w$). Then the mixture was ground in an agate mortar to a fine powder. The powder was pressed into a plate and the FTIR spectra were recorded by 40 scans in the range of 400 to 4000 $cm^{-1}$ with a resolution of 4 $cm^{-1}$ [17].

### 2.8. Low-Field Nuclear Magnetic Resonance (LF-NMR) of Wheat Dough

The quantitative determination of water structure in dough using LF-NMR was carried out with a 22.4 MHz NMR Analyzer PQ001 (Niumag Co., Ltd., Shanghai, China). The 90° and 180° pulse length was firstly measured to obtain the spin–spin relaxation time ($T_2$) by a CPMG pulse sequence. A total of 1000 echoes were collected. The pulse interval was 0.08 ms, the cycle delay was 1.5 s and 16 scans were collected.

### 2.9. Farinographic Property Determination of Wheat Dough

This property was determined according to the AACC Method 54-21 (2000) by a farinograph (Brabender Technologie GmbH & Co. KG, Duisburg, Germany). Farinographic properties, including water absorption, development time, stability time and degree of softening were recorded.

### 2.10. Dynamic Viscoelastic Property Determination of Wheat Dough

A quantity of 50 g dough mixed with or without wheat bran was kneaded by the Doughlab system (Perten Instruments, Hägersten, Sweden) at 65 r/min for 10 min. Afterwards, the mixture was maintained at 25 °C for 0.5 h to obtain test samples. Dynamic viscoelastic properties of dough with wheat bran were measured by a HAAKE MARS-III rheometer (Thermo Fisher Scientific Inc., Sunny-vale, CA, USA). A frequency sweep test on the dough was performed using a steel plate with a diameter of 20 mm gapped by 1 mm.

### 2.11. Statistical Analysis

Unless otherwise specified, 3 independent trials were conducted, and values are presented as the arithmetic means ± standard deviation (SD). Statistical differences were evaluated using ANOVA and Duncan's multiple range test. Values were considered to be significantly different at $p < 0.05$ (SPSS for Window 18.0).

## 3. Results and Discussion

### 3.1. Physicochemical Properties of Extruded Wheat Bran

#### 3.1.1. Component Analysis

The basic components of extruded wheat bran are shown in Figure 1a. Protein, starch, lipids, TDF, SDF and IDF contents of raw wheat bran were 18.37 ± 0.12%, 33.38 ± 1.37%, 4.08 ± 0.24%, 41.29 ± 2.89%, 3.10 ± 0.22% and 38.28 ± 1.37%, respectively. After extrusion, there was no significant difference observed in TDF and protein contents. The yield of SDF was increased to 7.65%, and the yields of IDF and starch were decreased to 33.31% and 30.57%, respectively. The increased content of SDF and decreased content of IDF suggested the conversion of IDF to SDF by extrusion. This was due to the disruption of non-covalent or covalent bonds in IDF, resulting in the production of soluble small-molecular-weight compounds accompanied by the change of molecular polarity [12]. Our previous study found that the content of dietary fiber components, for instance, arabinoxylan and $\beta$-glucan, and the proportion of high molecular weight granules of SDF were generally increased during extrusion [20,21]. According to reports, the influence of extrusion on TDF content was less than that of boiling, autoclaving or steaming [22]. Some researchers proposed that the unchanged contents of TDF were mainly attributed to the simultaneous formation of resistant starch at the same time as degradation of dietary fiber [5], but since starch content was decreased ($p \leq 0.01$), this was not the case in this study. The reason for the lowered starch content was the force degradation during extrusion, resulting in the cleavage of intermolecular hydrogen bonds and the degradation of macromolecules in starch. Additionally, there would be some hydration and gelatinization of the starch during extrusion [7,13,14]. The protein content of wheat bran after extrusion was almost unchanged, but extrusion has been reported to cause the hydration and denaturation of proteins, which in turn affected their biological activity or potency, such as the improvement of intrinsic digestibility, the reduction of trypsin inhibitor activity and drops in serum and liver cholesterol levels [7]. Therefore, extrusion treatment changed the main composition of wheat bran, greatly promoting the conversion of IDF to SDF.

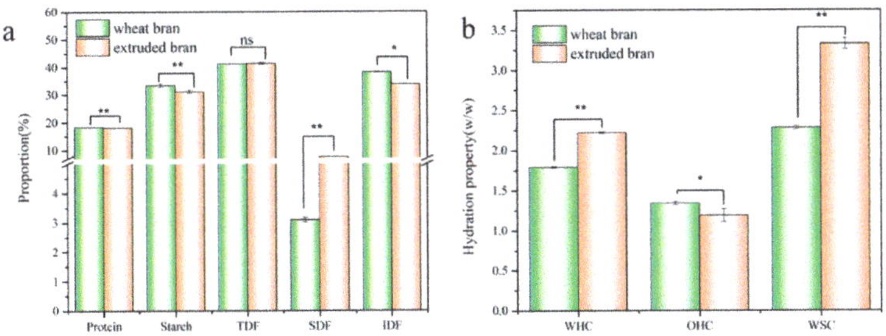


**Figure 1.** Contents of protein, starch, TDF, SDF and IDF of wheat bran (**a**); WHC, OHC and WSC value (**b**). Compared with raw bran, ns: no significant difference; *: significant difference (*: $p \leq 0.05$; **: $p \leq 0.01$).

### 3.1.2. Swelling Capacity and Solvent Retention Capacity

Studies have reported that the water-related properties, oil extraction capacity and swelling capacity of dietary fiber materials in functional foods were closely related to the rheological and functional properties. WHC, OHC and WSC of wheat bran before and after extrusion are illustrated in Figure 1b. In particular, WHC improved from 1.79% to 2.22%, and WSC increased from 2.28% to 3.33% while OHC deteriorated by 11.19%. As illustrated above, the degradation of SDF and some materials with low polymerization degree and high water-binding ability would be prompted by extrusion. There was more exposure of hydrophilic groups, which improved the water-binding ability of wheat bran and thus increased the WHC and reduced the OHC. Moreover, the multiple forces including mixing, stirring, shearing and puffing in extrusion process led to a pronounced change in the three-dimensional structure of wheat bran. The produced porous and loose structure was proportional to the improved retention and swelling ability of water, thus leading to the increase of WHC and WSC. High WHC was positively correlated with good hydration properties. The high hydration performance could facilitate water absorption of dietary fibers and their expansion to a gelation form, thereby increasing the food viscosity and delaying or preventing the excess cholesterol absorption in food. Additionally, the excellent hydrophilic properties could promote bowel movements and reduce risks of constipation, diverticulosis and colorectal cancer [23–25].

### 3.1.3. Morphological Characterization

The microstructures of unextruded and extruded wheat bran are displayed in Figure 2. The structure of the raw sample was dense. The surface was smooth and had no obvious pores, factures or faults. After extrusion, the surface of bran became loose, porous and rough, with some small holes and obvious fragments or faults on it. During the extrusion process of wheat bran, based on the principle of gas phase change and thermal pressing effect, the high shearing action between screw and barrel as well as the heating action outside the sleeve jointly produced the internal pressure in the barrel. Under such conditions of high temperature, pressure and shear force, wheat bran changed from powdery or granular form to pasty form. In particular, starch was gelatinized and cracked, protein was denatured and recombined and fiber was partially degraded and refined, resulting in a melted state of the bran micro-surface [9,21]. Then, under the strong pressure difference, occurring at the outlet of the extruder, the moisture contained in the bran was instantaneously vaporized and flashed, so that the bran was expanded to form a loose and puffed structure. Other studies also confirmed that the extrusion process was prone to generate the above changes in microstructure and morphology of bran, which was beneficial to be hydrolyzed by enzymes [26]. Such micro-structural deformation should be attributed to the fast and full dissolution of SDF and good hydration features of bran.

**Figure 2.** SEM graphs of raw (**a**) and extruded bran (**b–d**) powders.

### 3.2. Water States of Wheat Dough

Water distribution and mobility in dough have significant influence on the rheological behavior and technical features. Water, either bound to flour ingredients or free, is not homogeneously distributed in dough. To explore the water-migration performance and competitive water sequestration existed between wheat bran and gluten matrix, we used LF-NMR to determine the $T_2$ relaxation times and water contents of wheat dough (Figure 3). There are two CPMG proton populations, $T_{21}$ and $T_{22}$, which stand for bound water and free water, respectively. The relaxation time stands for the water mobility and peak area reflects the relative amount of H-bonds. A long relaxation time indicates a high degree of water freedom. For wheat dough, it has been inferred that $T_{21}$ represents the overlapping population of water on the surface of starch granules and water surrounding the gluten strands [27]. According to some researchers, bound water is thought to facilitate gluten formation and contribute to the supramolecular organization of dough structure [15]. $T_{22}$ is the capillary water between gluten network and wheat bran, which is related to the flow properties of the dough [3,15].

As shown in Table 2, with the increase in gluten content, $T_{21}$ relaxation time and $A_{21}$ peak-area ratio were increased. It indicated that the increase of gluten addition increased the amount of water hydrogen-bonded to the polymer matrix while weakening the water affinity. The increased $A_{21}$ peak-area ratio represents the relative content of bound water was increased in dough system. For flour dough with the addition of raw bran (RWD groups), $T_{21}$ relaxation time decreased in company with the increase of $A_{21}$ population and $T_{22}$ relaxation time. Since wheat bran had a stronger water-binding ability than gluten network, the increased $A_{21}$ population in RWD groups possibly contained the water absorbed by wheat bran plus intra-granular water in starch. Furthermore, the left shift of $T_{21}$ peak implied that bran sequestered bound water and prevented its free migration. There was no effect on the free water amount as evidenced from there being no regular changes in $A_{22}$ population. The above observations were in agreement with the fact that wheat bran not only sequestered water but also affected water activity.

For flour dough with the addition of extruded bran (EWD groups), Table 2 shows that $T_{21}$ relaxation times became shorter and $A_{21}$ populations tended to be larger, and $T_{22}$ peaked with the opposite changing trend compared to RWD groups. This phenomenon demonstrated higher populations of more tightly bound water and less populations of free water in EWD groups. As analyzed above, extruded bran contained more SDF with greater hydrophilic groups. Both the exposure of binding sites in bran structure and the damage of starch by extruding treatment should be responsible for the increased $A_{21}$ population. Because of the higher WHC of extruded bran than gluten proteins, water molecules tended to move from free water to bound water as presented by the shortened $T_{21}$ relaxation time. It has been recognized that hydration is supposed to be the important process for protein polymerization and dough formation. Water could act as a solvent that dissolves water-soluble proteins with low molecular weight and induce conformational changes in hydrophobic interactions [15]. Resulting from the partitioning of water between gluten and bran, it is hypothesized that extruded bran sequestered water, causing the partial dehydration of gluten matrix, thus limiting the water participating in the gluten formation process. Furthermore, the amount of extruded bran affected water retention capacity and dough rheology. To verify the hypothesis, the protein secondary structure was characterized for exploring the relationship between water sequestering and gluten development.

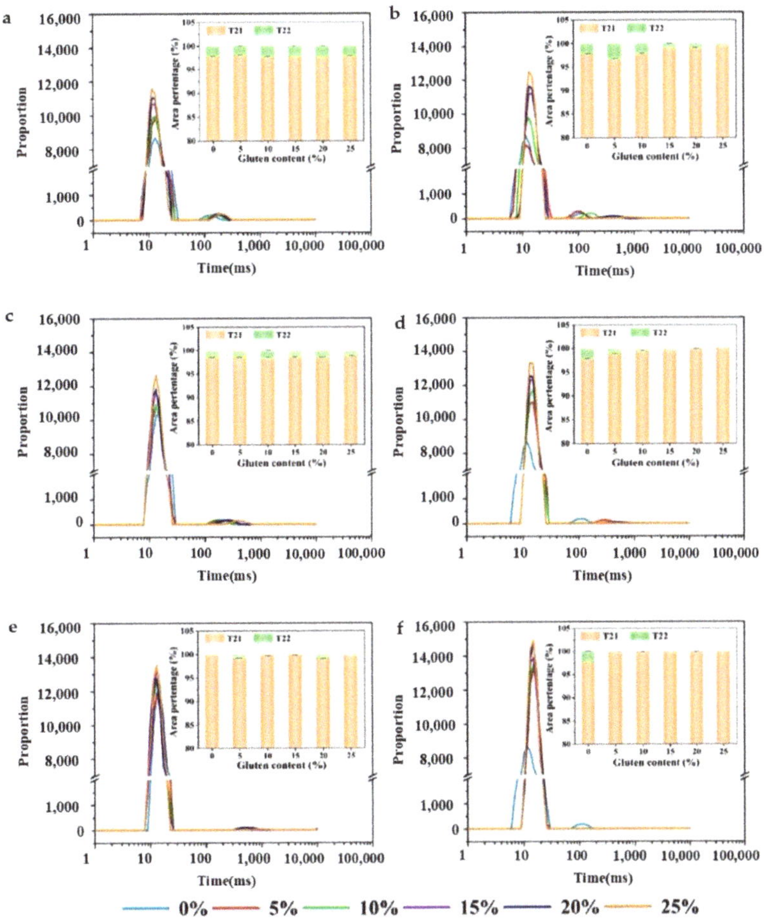


**Figure 3.** Distributed $T_2$ relaxation plots and water content for low- (**a**,**b**), middle- (**c**,**d**) and high- (**e**,**f**) gluten dough with incorporation of raw (**a**,**c**,**e**) and extruded bran (**b**,**d**,**e**).

**Table 2.** $T_2$ relaxation time and peak-area ratio of wheat dough.

| | Bran Addition/% | Raw Bran | | | | Extruded Bran | | | |
|---|---|---|---|---|---|---|---|---|---|
| | | $T_{21}$/ms | $T_{22}$/ms | $A_{21}$/% | $A_{22}$/% | $T_{21}$/ms | $T_{22}$/ms | $A_{21}$/% | $A_{22}$/% |
| High-gluten flour | 0 | $15.21 \pm 0.10^{a}$ | $618.18 \pm 0.00^{d}$ | $99.87 \pm 0.00^{a}$ | $0.10 \pm 0.00^{d}$ | – | – | – | – |
| | 5 | $13.23 \pm 0.00^{b}$ | $444.78 \pm 0.00^{f}$ | $99.07 \pm 0.00^{cd}$ | $0.93 \pm 0.00^{b}$ | $15.11 \pm 0.00$ | – | $100.00 \pm 0.00$ | – |
| | 10 | $13.23 \pm 0.00^{b}$ | $920.05 \pm 0.00^{b}$ | $99.75 \pm 0.01^{ab}$ | $0.17 \pm 0.01^{d}$ | $15.11 \pm 0.00$ | – | $100.00 \pm 0.00$ | – |
| | 15 | $13.23 \pm 0.00^{b}$ | $1344.94 \pm 0.39^{a}$ | $99.82 \pm 0.23^{ab}$ | $0.08 \pm 0.02^{d}$ | $15.11 \pm 0.00$ | – | $100.00 \pm 0.00$ | – |
| | 20 | $12.14 \pm 0.55^{c}$ | $496.63 \pm 0.22^{e}$ | $98.91 \pm 0.14^{d}$ | $1.09 \pm 0.01^{a}$ | $15.11 \pm 0.00$ | – | $100.00 \pm 0.00$ | – |
| | 25 | $13.23 \pm 0.00^{b}$ | $777.22 \pm 0.71^{c}$ | $99.44 \pm 0.07^{bc}$ | $0.55 \pm 0.08^{c}$ | $15.11 \pm 0.00$ | – | $100.00 \pm 0.00$ | – |
| Middle-gluten flour | 0 | $13.22 \pm 0.02$ | $214.09 \pm 0.00^{d}$ | $98.80 \pm 0.02^{a}$ | $1.20 \pm 0.02^{e}$ | – | – | – | – |
| | 5 | $13.23 \pm 0.00$ | $187.51 \pm 0.00^{e}$ | $98.51 \pm 0.03^{b}$ | $1.49 \pm 0.03^{c}$ | $15.11 \pm 0.00^{a}$ | $279.07 \pm 0.00^{c}$ | $98.90 \pm 0.01^{e}$ | $1.09 \pm 0.02^{a}$ |
| | 10 | $13.23 \pm 0.00$ | $171.99 \pm 0.77^{d}$ | $98.20 \pm 0.02^{d}$ | $1.80 \pm 0.02^{a}$ | $15.11 \pm 0.00^{a}$ | $415.35 \pm 0.00^{b}$ | $99.22 \pm 0.05^{d}$ | $0.78 \pm 0.05^{b}$ |
| | 15 | $13.23 \pm 0.00$ | $255.98 \pm 1.15^{b}$ | $98.56 \pm 0.03^{b}$ | $1.44 \pm 0.03^{c}$ | $15.11 \pm 0.00^{a}$ | $474.22 \pm 2.24^{b}$ | $99.38 \pm 0.05^{c}$ | $0.62 \pm 0.05^{c}$ |
| | 20 | $13.23 \pm 0.00$ | $244.43 \pm 0.00^{c}$ | $98.38 \pm 0.03^{c}$ | $1.62 \pm 0.03^{b}$ | $13.23 \pm 0.00^{b}$ | $705.80 \pm 1.89^{a}$ | $99.76 \pm 0.03^{b}$ | $0.18 \pm 0.03^{d}$ |
| | 25 | $13.23 \pm 0.00$ | $365.92 \pm 2.79^{a}$ | $98.70 \pm 0.04^{a}$ | $1.30 \pm 0.03^{d}$ | $13.23 \pm 0.63^{b}$ | – | $100.00 \pm 0.00^{a}$ | – |
| Low-gluten flour | 0 | $13.22 \pm 0.02^{a}$ | $125.99 \pm 0.00^{f}$ | $98.50 \pm 0.02$ | $1.50 \pm 0.02$ | – | – | – | – |
| | 5 | $13.23 \pm 0.00^{a}$ | $157.43 \pm 0.68^{d}$ | $98.50 \pm 0.04$ | $1.50 \pm 0.04$ | $11.59 \pm 0.00$ | $96.65 \pm 4.57^{c}$ | $97.84 \pm 0.02^{d}$ | $2.16 \pm 0.02^{a}$ |
| | 10 | $13.23 \pm 0.00^{a}$ | $143.84 \pm 0.00^{e}$ | $98.20 \pm 0.04$ | $1.80 \pm 0.04$ | $13.23 \pm 0.00$ | $164.23 \pm 1.76^{c}$ | $98.43 \pm 0.03^{c}$ | $1.57 \pm 0.03^{b}$ |
| | 15 | $12.68 \pm 0.55^{ab}$ | $179.75 \pm 0.77^{c}$ | $98.48 \pm 0.26$ | $1.52 \pm 0.26$ | $15.11 \pm 0.00$ | $415.35 \pm 7.21^{b}$ | $98.96 \pm 0.00^{b}$ | $1.04 \pm 0.00^{c}$ |
| | 20 | $12.14 \pm 0.55^{ab}$ | $187.51 \pm 0.00^{b}$ | $98.28 \pm 0.03$ | $1.72 \pm 0.03$ | $13.23 \pm 0.00$ | $415.35 \pm 1.96^{b}$ | $99.02 \pm 0.02^{b}$ | $0.98 \pm 0.02^{c}$ |
| | 25 | $11.59 \pm 0.00^{b}$ | $196.37 \pm 0.89^{a}$ | $98.24 \pm 0.02$ | $1.76 \pm 0.02$ | $13.23 \pm 0.00$ | $805.83 \pm 5.42^{a}$ | $99.50 \pm 0.13^{a}$ | $0.50 \pm 0.05^{d}$ |

Values (mean ± SD, $n = 3$) with different letters have mean values that are significantly different ($p < 0.05$).

### 3.3. Protein Secondary Structure of Wheat Dough

The FTIR spectroscopy was performed to study second structure changes in gluten with the incorporation of wheat bran. The contents of α-helix, β-sheet, β-turn and random structure were calculated from relative peak areas of 1652~1660 $cm^{-1}$, 1660~1685 $cm^{-1}$, 1620~1644 $cm^{-1}$ and 1644~1652 $cm^{-1}$, respectively [17].

In Figure 4, RWD groups show that β-turn and β-sheet were major protein secondary structures, which occupied 32.23~39.99% and 28.22~33.26%, respectively, while α-helix and random accounted for 15.63~19.23% and 13.73~16.94%, respectively. The increase of gluten content increased the β-turn content with no change in the other three structure contents. As Bock et al. reported, in the model gluten-only dough, gluten predominantly exists as β-turn structures. So, with the increment of gluten content in dough samples, the β-turn content increased. According to their study, these structures in wheat dough could have a relationship with β-spiral domains in glutenin polypeptides. After extrusion treatment, there are many changes in protein secondary structures of gluten in EWD groups. The β-turn contents of EWD groups were higher than those of RWD groups, while the β-sheet content presented the opposite tendency. With the increase of extruded bran content, β-turn content reduced, and β-sheet content increased for low-gluten and middle-gluten flour dough. It was suggested that the dramatic increase of β-sheet was at the expense of the decreased content of β-turn. However, for EWD groups at high-gluten content dough, the conversion of β-turn to β-sheet was not apparent. Some reports have pointed out that β-turn was the preferred structure of gluten in the fully hydrated state [28]. The mechanistic explanation of these empirical observations should be attributed to the redistribution of moisture between extruded bran and gluten in wheat dough as evidenced from the above section. The water-sequestering capability of extruded bran promoted the binding ability of dough to bound water as evidenced by the increased $A_{21}$ and decreased $T_{21}$. The data also showed that the bran could compete with the gluten network for the bound water, which in turn caused protein conformational variations [3,15]. However, the transformation was minimal in higher-gluten dough due to sufficient water being available in the system to adequately hydrate the dough system. The results clearly indicated that extruded bran addition influenced the direct linkage between trans-conformational changes in gluten and gluten-water interaction extent in dough, and the degree of this transformation must be a function of the ratio of extruded bran to gluten content.

In fact, the protein secondary structures are closely associated with the gluten network formation and the gluten strength. According to reports, the β-sheet is thought to be the most stable of protein conformation [3], and the high degree of protein polymerization is strongly related to the buildup of the stable protein secondary structures [15]. Therefore, in the lower-gluten dough, the higher proportion of β-sheet structure should demonstrate the higher ordered structure and facilitate the polymerization of protein in EWD groups than RWD groups. Addition of extruded bran made the gluten tend to adopt a β-sheet configuration, resulting from the water redistribution and gluten dehydration. This structural transformation made the protein aggregate firmly to form a three-dimensional viscoelastic network. The loss of β-turn structures and the formation of β-sheet structures, thus, implied the formation of a more elastic gluten network. Other studies also showed the reconstruction of protein secondary structures of wheat dough by bran addition. Such a change was considered to be one of the structural elements responsible for dough viscoelasticity [3,21].

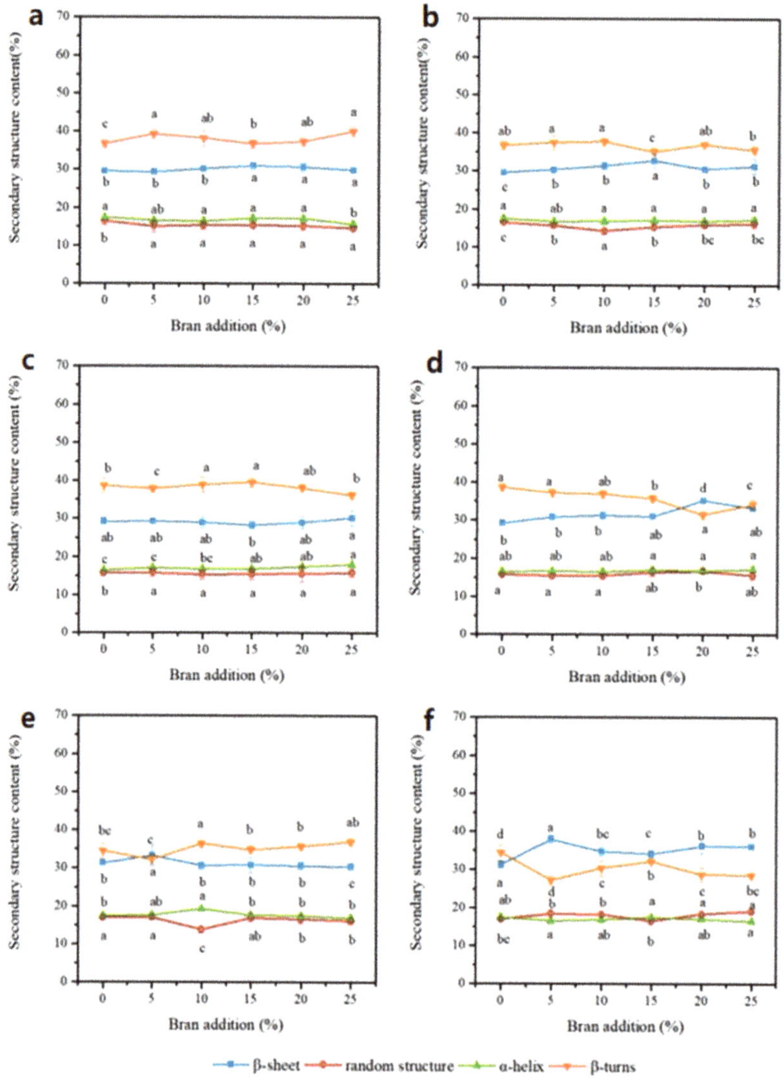

**Figure 4.** Protein secondary structure contents of wheat dough at high- (**a**,**b**), middle- (**c**,**d**), low- (**e**,**f**) gluten dough with raw (**a**,**c**,**e**) and extruded bran (**b**,**d**,**f**). Dots (mean ± SD, $n = 3$) with different letters have mean values that are significantly different ($p < 0.05$).

### 3.4. Rheological Properties of Wheat Dough

#### 3.4.1. Farinographic Property

In Table 3, water absorption (WA) of all samples showed a gradually increasing trend with the addition of bran, and EWD groups caused a further increase in WA compared to RWD groups. It was typically attributed to the high WHC of extruded sample and its strong water competition capacity. The high presence of hydrophilic groups in SDF, the damage of starch (the starch content in wheat bran was decreased from 33.38 ± 1.37% to 30.01 ± 0.49% after extrusion treatment) and the exposure of binding sites in extruded bran with reduced particle sizes could lead to the higher WA in EWD groups. The results are in agreement with the reports of Gómez et al. [9] and Wang et al. [29], which found that bran addition caused an increase in the of WA of dough. Dough development time (DT) and stability value (ST) stand for the flour strength, with higher values correlating with the stronger doughs [29]. With regard to DT, bran addition raised the DT value for low-gluten dough, which reflected the increased time to reach the maximum dough consistency, and EWD groups showed greater increased tendency. However, with the increment of gluten content, the DT value was reduced with bran addition. With regard to ST, some researchers have found that the addition of bran deteriorated the dough stability, while other studies provided opposite results [9,29]. Our results showed that ST was prolonged for low-gluten flour while it was shortened at higher-gluten conditions with the increase of bran amount. The increased DT and ST in low-gluten flour may be caused by the interference of bran on the gluten formation process caused by the water competition effect, which prolonged the dough-development process and improved the stability. Moreover, the improved stability upon bran addition may also be related to the increased rigidity of dough. For higher-gluten flours, bran possessed a weaker ability to competing for water owing to the less exposed binding sites within a relatively intensive gluten network. This is because bran influenced gluten development by the formation of steric hindrance, or by the dilution of the gluten or by piercing the gas cells, which reduced the gluten connectivity and thus lowered the DT and stabilities. Furthermore, the negative influences were enhanced when extruded bran was added. Possible explanations might be that the powered hydrophilic micro-environment, the liberation of soluble reactive components and the increased particle surface of extruded bran contribute to easier interactions of gluten with reaction components. According to Noort et al. [30], ferulic acid monomer bound to the cell wall could interact with gluten proteins in particular.

The mixing tolerance index, Ds, was used to assess the softening degree of the dough [3]. The Ds results showed that, in the low-gluten system, raw bran addition improved the hardness of the dough sheet, while the addition of extruded bran further promoted this tendency. Thus, farinographic results suggested that extruded bran under low-gluten conditions had a positive influence on the strength of the gluten network.

**Table 3.** Farinographic properties of wheat flours.

| Flour Type | Bran Addition /% | Untreated Bran | | | | Extruded Bran | | | |
|---|---|---|---|---|---|---|---|---|---|
| | | Water Absorption (WA) /% | Development Time (DT) /min | Stability Time (ST) /min | Degree of Softening (Ds) /FU | Water Absorption (WA) /% | Development Time (DT) /min | Stability Time (ST) /min | Degree of Softening (Ds) /FU |
| High-gluten flour | 0 | $62.9 \pm 0.0^{f}$ | $15.0 \pm 0.1^{a}$ | $19.6 \pm 0.1^{a}$ | – | $63.1 \pm 0.0^{c}$ | $15.0 \pm 0.1^{a}$ | $19.6 \pm 0.0^{a}$ | – |
| | 5 | $64.4 \pm 0.1^{e}$ | $9.1 \pm 0.0^{c}$ | $17.1 \pm 0.0^{b}$ | – | $65.3 \pm 0.1^{c}$ | $13.7 \pm 0.2^{b}$ | $14.1 \pm 0.2^{b}$ | – |
| | 10 | $65.6 \pm 0.0^{d}$ | $10.1 \pm 0.1^{b}$ | $15.5 \pm 0.1^{c}$ | – | $68.5 \pm 0.1^{b}$ | $11.1 \pm 0.3^{c}$ | $14.6 \pm 0.2^{b}$ | – |
| | 15 | $68.7 \pm 0.0^{c}$ | $9.0 \pm 0.0^{c}$ | $15.7 \pm 0.2^{c}$ | – | $72.7 \pm 0.1^{b}$ | $8.8 \pm 0.1^{d}$ | $11.4 \pm 0.4^{c}$ | – |
| | 20 | $70.0 \pm 0.2^{b}$ | $7.2 \pm 0.2^{e}$ | $13.4 \pm 0.3^{d}$ | – | $76.2 \pm 0.3^{a}$ | $8.4 \pm 0.1^{d}$ | $7.8 \pm 0.2^{d}$ | – |
| | 25 | $74.3 \pm 0.1^{a}$ | $7.8 \pm 0.1^{d}$ | $11.2 \pm 0.2^{e}$ | – | $77.3 \pm 0.2^{a}$ | $7.5 \pm 0.2^{e}$ | $6.4 \pm 0.1^{d}$ | $112.8 \pm 7.3$ |
| Middle-gluten flour | 0 | $58.8 \pm 0.1^{f}$ | $5.3 \pm 0.3^{a}$ | $12.6 \pm 0.2^{a}$ | $45.0 \pm 1.2^{c}$ | $58.8 \pm 0.2^{e}$ | $5.2 \pm 0.1^{a}$ | $13.4 \pm 0.1^{a}$ | $44.4 \pm 2.1^{e}$ |
| | 5 | $60.9 \pm 0.1^{e}$ | $3.2 \pm 0.1^{c}$ | $7.8 \pm 0.1^{b}$ | $93.5 \pm 2.0^{b}$ | $62.0 \pm 0.0^{d}$ | $2.5 \pm 0.0^{d}$ | $7.3 \pm 0.1^{b}$ | $72.4 \pm 3.2^{d}$ |
| | 10 | $62.8 \pm 0.2^{d}$ | $3.3 \pm 0.2^{c}$ | $6.8 \pm 0.1^{c}$ | $103.2 \pm 1.7^{b}$ | $65.6 \pm 0.1^{c}$ | $3.4 \pm 0.2^{c}$ | $4.8 \pm 0.2^{c}$ | $108.6 \pm 9.0^{c}$ |
| | 15 | $65.4 \pm 0.1^{c}$ | $4.2 \pm 0.1^{b}$ | $6.1 \pm 0.3^{d}$ | $116.8 \pm 1.5^{a}$ | $70.0 \pm 0.2^{b}$ | $3.8 \pm 0.2^{b}$ | $4.1 \pm 0.1^{cd}$ | $122.9 \pm 6.4^{bc}$ |
| | 20 | $68.7 \pm 0.2^{b}$ | $3.8 \pm 0.3^{b}$ | $5.5 \pm 0.2^{d}$ | $129.1 \pm 3.5^{a}$ | $73.0 \pm 0.3^{a}$ | $4.0 \pm 0.0$ | $3.7 \pm 0.0^{d}$ | $136.4 \pm 4.8^{a}$ |
| | 25 | $73.1 \pm 0.1^{a}$ | $5.3 \pm 0.2^{a}$ | $6.3 \pm 0.2^{d}$ | $128.9 \pm 2.4^{a}$ | $75.1 \pm 0.1^{a}$ | $4.0 \pm 0.1^{b}$ | $2.8 \pm 0.1^{e}$ | $135.4 \pm 5.1^{a}$ |
| Low-gluten flour | 0 | $49.3 \pm 0.1^{e}$ | $1.2 \pm 0.0^{d}$ | $1.1 \pm 0.1^{e}$ | $132.1 \pm 2.1^{a}$ | $49.3 \pm 0.1^{e}$ | $1.1 \pm 0.3^{d}$ | $1.1 \pm 0.3^{d}$ | $132.1 \pm 2.1^{a}$ |
| | 5 | $49.6 \pm 0.1^{e}$ | $1.2 \pm 0.0^{d}$ | $1.4 \pm 0.0^{e}$ | $119.8 \pm 4.1^{b}$ | $49.3 \pm 0.0^{e}$ | $1.3 \pm 0.1^{d}$ | $1.7 \pm 0.2^{d}$ | $126.0 \pm 3.3^{b}$ |
| | 10 | $52.5 \pm 0.0^{d}$ | $1.3 \pm 0.1^{cd}$ | $2.7 \pm 0.2^{d}$ | $110.2 \pm 1.9^{c}$ | $52.5 \pm 0.1^{d}$ | $1.5 \pm 0.0^{d}$ | $3.2 \pm 0.4^{c}$ | $121.7 \pm 5.0^{b}$ |
| | 15 | $57.4 \pm 0.2^{c}$ | $1.5 \pm 0.1^{d}$ | $3.4 \pm 0.1^{c}$ | $106.6 \pm 2.3^{c}$ | $56.1 \pm 0.3^{c}$ | $3.9 \pm 0.2^{c}$ | $5.4 \pm 0.2^{b}$ | $95.4 \pm 2.4^{cd}$ |
| | 20 | $60.7 \pm 0.0^{b}$ | $1.8 \pm 0.0^{b}$ | $5.5 \pm 0.2^{b}$ | $90.3 \pm 0.7^{d}$ | $58.2 \pm 0.2^{bc}$ | $5.3 \pm 0.3^{b}$ | $4.5 \pm 0.2^{bc}$ | $86.2 \pm 7.9^{d}$ |
| | 25 | $62.5 \pm 0.1^{a}$ | $3.9 \pm 0.1^{a}$ | $6.4 \pm 0.1^{a}$ | $98.0 \pm 1.4^{d}$ | $63.5 \pm 0.1^{a}$ | $6.7 \pm 0.2^{a}$ | $7.0 \pm 0.0^{a}$ | $45.9 \pm 3.7^{e}$ |

Values (mean ± SD, $n = 3$) with different letters have mean values that are significantly different ($p < 0.05$).

### 3.4.2. Dynamic Rheological Behavior

Figures 5 and 6 show the dynamic rheological properties of dough fortified with 0~25% wheat bran. In all dough samples, both the storage ($G'$) and loss ($G''$) moduli increased with the increasing frequency, which demonstrated their dependence on the frequency [31]. Generally, the dough samples demonstrated a response typical of a cross-linked polymer network mainly manifesting solid-like behavior as evidenced by $G'' < G'$ (loss tangent, tan$\delta$ < 1) [32]. $G'$ and $G''$ values of both RWD and EWD groups showed a rising tendency with the frequency increased from 0.1 to 10 Hz. This behavior has already been reported by many researchers [3,31]. A greater $G'$ normally indicates a more elastic and solid-like structure. As shown in Figures 5 and 6, EWD groups showed stronger mechanical strength than RWD groups. It suggested that the much stiffer doughs had lower flowing tendency. Compared with control dough, RWD groups showed lower tan$\delta$ values, and EWD groups enlarged the difference. Furthermore, tan$\delta$ decreased with the increased extruded bran. The increased rheological moduli and reduced loss tangent might be owing to the water competition of wheat bran with the gluten network in the mixing process, which restrained the water uptake of the gluten network and gave rise to interparticle interaction between water and other dough components in a strengthened dough system, finally contributing to its elastic properties. Ronda et al. [33,34] noted that bran prompted solid-like behavior of dough. On this basis, viscoelastic profiles indicated that EWD groups formed a more compact and less extensible gluten network than RWD groups. In the processing of flour products, such as mixing, stretching and fermentation, it is necessary to form a continuous gluten matrix to ensure the stability of the preprepared dough. However, the formation of an excessively compact gluten matrix could make doughs have poor processing adaptability, such as possibility of hard dough.

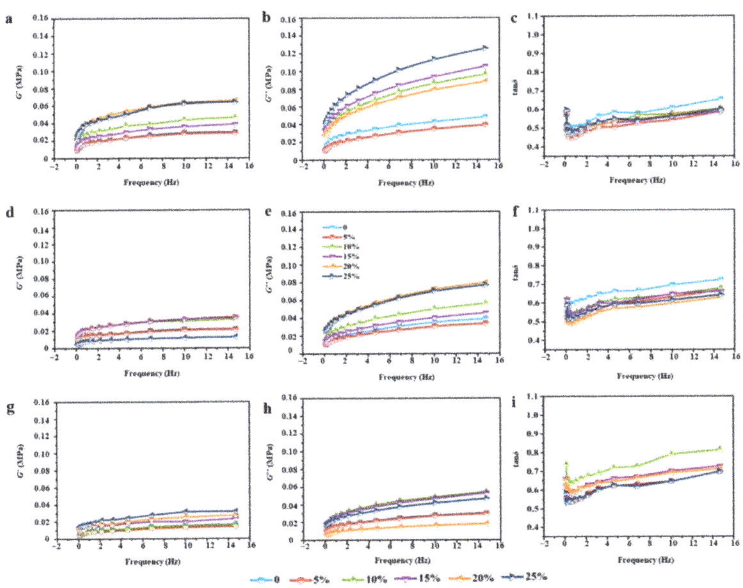


**Figure 5.** $G'$, $G''$ and tan$\delta$ of high-gluten (**a–c**), middle-gluten (**d–f**) and low-gluten (**g–i**) dough with raw bran.

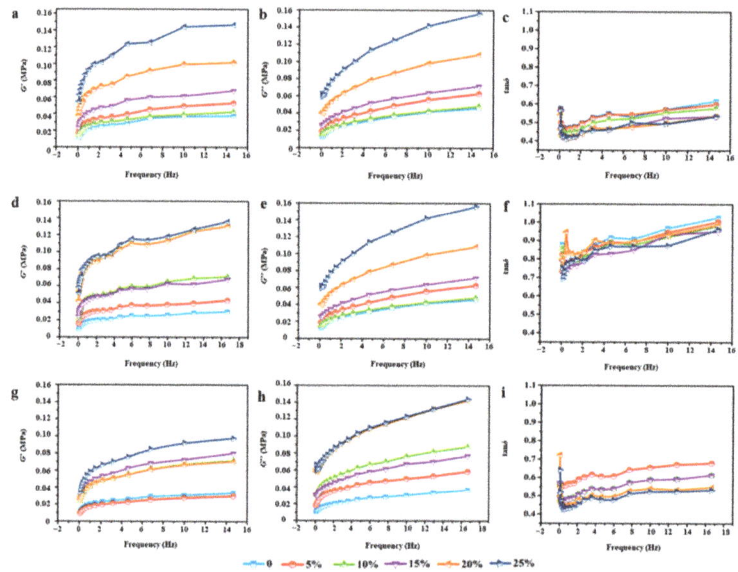

**Figure 6.** $G'$, $G''$ and tan$\delta$ of high-gluten (**a**–**c**), middle-gluten (**d**–**f**) and low-gluten (**g**–**i**) dough with extruded bran.

## 4. Conclusions

The morphological and compositional characteristics, swelling and solvent retention capacity of wheat bran treated by extrusion were analyzed. These experiments confirmed that the solubility of dietary fiber was increased by a factor of 1.47 and the porous structure of bran was formed after extrusion processing. Thus the consequent hydration and swelling capacities were higher in extruded bran. The effect of extruded wheat bran on the structural and rheological properties of wheat dough with different gluten contents was elucidated. The water state and protein secondary structure of gluten dough was studied to reveal the underlying physical mechanism by which bran–gluten interaction impacted dough properties. LF-NMR results revealed that there were two distinct water populations formed in the gluten dough. Addition of extruded bran caused water redistribution, as presented by the larger proportion of tightly bound water to the gluten matrix. The bran–gluten interaction influenced by water sequestering affected the protein secondary structure as evidenced by the conversion of $\beta$-turn into $\beta$-sheet especially in lower-gluten dough, which indicated the partial dehydration of gluten and consequently formed a more stable and polymerized gluten network. Farinographic data showed that extruded bran increased DT and ST by factors of 2.25 and 4.82, respectively, and decreased the Ds by 25.81% only at lower-gluten content of dough. Viscoelastic data confirmed that extruded bran increased $G'$, $G''$ and reduced tan$\delta$ of wheat dough at all gluten contents, indicating that the incorporation of extruded bran would improve the dough elasticity. These results would be useful for interpreting the modification effects and mechanism of extrusion on cereal brans and guide the application of the extrusion technique on bran processing and its flour products development.

**Author Contributions:** Writing—original draft, R.L. (Ranran Li) and C.W.; Writing—review and editing, R.L. (Ranran Li) and C.W.; Investigation, X.X. and Y.W.; Supervision, W.S., R.L. (Rui Liu) and T.W.; Funding acquisition, M.Z. and W.S. All authors have read and agreed to the published version of the manuscript.

**Funding:** The work is supported by National Natural Science Foundation of China (32172169) and Tianjin Natural Science Foundation (22JCZDJC00360).

**Data Availability Statement:** Data is contained within the article.

**Conflicts of Interest:** The authors declare no conflict of interest.

## References


1. Roye, C.; Henrion, M.; Chanvrier, H.; De Roeck, K.; De Bondt, Y.; Liberloo, I.; King, R.; Courtin, C.M. Extrusion-cooking modifies physicochemical and nutrition-related properties of wheat bran. *Foods* **2020**, *9*, 738. [CrossRef] [PubMed]
2. Zhang, S.; Jia, X.; Xu, L.; Xue, Y.; Pan, Q.; Shen, W.; Wang, Z. Effect of extrusion and semi-solid enzymatic hydrolysis modifications on the quality of wheat bran and steamed bread containing bran. *J. Cereal Sci.* **2022**, *108*, 103577. [CrossRef]
3. Sui, W.J.; Xie, X.; Liu, R.; Wu, T.; Zhang, M. Effect of wheat bran modification by steam explosion on structural characteristics and rheological properties of wheat flour dough. *Food Hydrocoll.* **2018**, *84*, 571–580. [CrossRef]
4. Khanpit, V.V.; Tajane, S.P.; Mandavgane, S.A. Extrusion for soluble dietary fiber concentrate: Critical overview on effect of process parameters on physicochemical, nutritional, and biological properties. *Food Res. Int.* **2022**, 1–22. [CrossRef]
5. Andersson, A.A.M.; Andersson, R.; Jonsäll, A.; Andersson, J.; Fredriksson, H. Effect of different extrusion parameters on dietary fiber in wheat bran and rye bran. *J. Food Sci.* **2017**, *82*, 1344–1350. [CrossRef]
6. Yan, X.G.; Ye, R.; Chen, Y. Blasting extrusion processing: The increase of soluble dietary fiber content and extraction of soluble-fiber polysaccharides from wheat bran. *Food Chem.* **2015**, *180*, 106–115. [CrossRef]
7. Korhonen, H.; Pihlanto-Leppäla, A.; Rantamäki, P.; Tupasela, T. Impact of processing on bioactive proteins and peptides. *Trends Food Sci. Technol.* **1998**, *9*, 307–319. [CrossRef]
8. Nawrocka, A.; Szymańska-Chargot, M.; Miś, A.; Wilczewska, A.Z.; Markiewicz, K.H. Dietary Fiber-Induced Changes in the Structure and Thermal Properties of Gluten Proteins Studied by Fourier Transform-Raman Spectroscopy and Thermogravimetry. *J. Agric. Food Chem.* **2016**, *64*, 094–2104. [CrossRef]
9. Gómez, M.; Jiménez, S.; Ruiz, E.; Oliete, B. Effect of extruded wheat bran on dough rheology and bread quality. *LWT Food Sci. Technol.* **2011**, *44*, 2231–2237. [CrossRef]
10. Sun, X.; Wu, S.; Koksel, F.; Xie, M.; Fang, Y. Effects of ingredient and processing conditions on the rheological properties of whole wheat flour dough during breadmaking—A review. *Food Hydrocolloid.* **2023**, *135*, 108123. [CrossRef]
11. Wu, Y.; Ye, G.; Li, X.; Wang, L.; Liu, Y.; Tan, B.; Shen, W.; Zhou, J. Comparison of quality characteristics of six reconstituted whole wheat flour with different modified bran. *LWT Food Sci. Technol.* **2022**, *153*, 112543. [CrossRef]
12. Menis-Henrique, M.E.C.; Scarton, M.; Piran, M.V.F.; Clerici, M.T.P.S. Cereal fiber: Extrusion modifications for food industry. *Curr. Opin. Food Sci.* **2020**, *33*, 141–148. [CrossRef]
13. Logié, N.; Della Valle, G.; Rolland-Sabaté, A.; Descamps, N.; Soulestin, J. How does temperature govern mechanisms of starch changes during extrusion. *Carbohydr. Polym.* **2018**, *184*, 57–65. [CrossRef] [PubMed]
14. Offiah, V.; Kontogiorgos, V.; Falade, K.O. Extrusion processing of raw food materials and by-products: A review. *Crit. Rev. Food Sci. Nutr.* **2019**, *59*, 2979–2998. [CrossRef] [PubMed]
15. Xiong, L.C.; Zhang, B.J.; Niu, M.; Zhao, S.M. Protein polymerization and water mobility in whole-wheat dough influenced by bran particle size distribution. *LWT Food Sci. Technol.* **2017**, *82*, 396–403. [CrossRef]
16. Bock, J.E.; Connelly, R.K.; Damodaran, S. Impact of bran addition on water properties and gluten secondary structure in wheat flour doughs studied by attenuated total reflectance Fourier transform infrared spectroscopy. *Cereal Chem.* **2013**, *90*, 377–386. [CrossRef]
17. Bock, J.E.; Damodaran, S. Bran-induced changes in water structure and gluten conformation in model gluten dough studied by Fourier transform infrared spectroscopy. *Food Hydrocoll.* **2013**, *31*, 146–155. [CrossRef]
18. Wang, P.; Jin, Z.Y.; Xu, X.M. Physicochemical alterations of wheat gluten proteins upon dough formation and frozen storage—A review from gluten, glutenin and gliadin perspectives. *Trends Food Sci. Technol.* **2015**, *46*, 189–198. [CrossRef]
19. Li, J.; Kang, J.; Wang, L.; Li, Z.; Wang, R.; Chen, Z.X.; Hou, G.G. Effect of water migration between arabinoxylans and gluten on baking quality of whole wheat bread detected by magnetic resonance imaging (MRI). *J. Agric. Food Chem.* **2012**, *60*, 6507–6514. [CrossRef]
20. Li, Q.; Liu, R.; Wu, T.; Zhang, M. Interactions between soluble dietary fibers and wheat gluten in dough studied by confocal laser scanning microscopy. *Food Res. Int.* **2017**, *95*, 19–27. [CrossRef]
21. Zhang, M.; Bai, X.; Zhang, Z.S. Extrusion process improves the functionality of soluble dietary fiber in oat bran. *J. Cereal Sci.* **2011**, *54*, 98–103. [CrossRef]
22. McCleary, B.; Prosky, L. *Advanced Dietary Fibre Technology*; John Wiley & Sons: New York, NY, USA, 2008.
23. Akalın, A.S.; Kesenkas, H.; Dinkci, N.; Unal, G.; Ozer, E.; Kınık, O. Enrichment of probiotic ice cream with different dietary fibers: Structural characteristics and culture viability. *J. Dairy Sci.* **2018**, *101*, 37–46. [CrossRef] [PubMed]
24. Wang, C.; Song, R.; Wei, S.; Wang, W.; Li, F.; Tang, X.; Li, N. Modification of insoluble dietary fiber from ginger residue through enzymatic treatments to improve its bioactive properties. *LWT Food Sci. Technol.* **2020**, *125*, 109220. [CrossRef]
25. Yan, J.; Hu, J.; Yang, R.; Zhang, Z.; Zhao, W. Innovative nanofibrillated cellulose from rice straw as dietary fiber for enhanced health benefits prepared by a green and scale production method. *ACS Sustain. Chem. Eng.* **2018**, *6*, 3481–3492. [CrossRef]
26. Haghighi-Manesh, S.; Azizi, M.H. Integrated extrusion-enzymatic treatment of corn bran for production of functional cake. *Food Sci. Nutr.* **2018**, *6*, 1870–1878. [CrossRef]

27. Lu, Z.; Seetharaman, K. $^{1}$H Nuclear magnetic resonance (NMR) and differential scanning calorimetry (DSC) studies of water mobility in dough systems containing barley flour. *Cereal Chem.* **2013**, *90*, 120–126. [CrossRef]
28. Arcila, J.A.; Weier, S.A.; Rose, D.J. Changes in dietary fiber fractions and gut microbial fermentation properties of wheat bran after extrusion and bread making. *Food Res. Int.* **2015**, *74*, 217–223. [CrossRef]
29. Wang, M.W.; Hamer, R.J.; van Vliet, T.; Oudgenoeg, G. Interaction of water extractable pentosans with gluten protein: Effect on dough properties and gluten quality. *J. Cereal Sci.* **2002**, *36*, 25–37. [CrossRef]
30. Noort, M.W.J.; van Haaster, D.; Hemery, Y.; Schols, H.A.; Hamer, R.J. The effect of particle size of wheat bran fractions on bread quality–Evidence for fibre–protein interactions. *J. Cereal Sci.* **2010**, *52*, 59–64. [CrossRef]
31. Mejia, C.D.; Gonzalez, D.C.; Mauer, L.J.; Campanella, O.H.; Hamaker, B.R. Increasing and stabilizing $\beta$-sheet structure of maize zein causes improvement in its rheological properties. *J. Agric. Food Chem.* **2012**, *60*, 2316–2321. [CrossRef]
32. Sarawong, C.; Schoenlechner, R.; Sekiguchi, K.; Berghofer, E.; Ng, P.K.W. Effect of extrusion cooking on the physicochemical properties, resistant starch, phenolic content and antioxidant capacities of green banana flour. *Food Chem.* **2014**, *143*, 33–39. [CrossRef] [PubMed]
33. Ronda, F.; Perez-Quirce, S.; Lazaridou, A.; Biliaderis, C.G. Effect of barley and oat $\beta$-glucan concentrates on gluten-free rice-based doughs and bread characteristics. *Food Hydrocoll.* **2015**, *48*, 197–207. [CrossRef]
34. Wang, L.; Ye, F.Y.; Li, S.; Wei, F.B.; Chen, J.F.; Zhao, G.H. Wheat flour enriched with oat $\beta$-glucan: A study of hydration, rheological and fermentation properties of dough. *J. Cereal Sci.* **2017**, *75*, 143–150. [CrossRef]

MDPI AG
Grosspeteranlage 5
4052 Basel
Switzerland
Tel.: +41 61 683 77 34

*Foods* Editorial Office
E-mail: foods@mdpi.com
www.mdpi.com/journal/foods

www.ingramcontent.com/pod-product-compliance
Lightning Source LLC
Chambersburg PA
CBHW062045220926
51850CB00010B/2671
* 9 7 8 3 7 2 5 8 7 6 4 5 7 *